THE FRONTIERS COLLECTION

T0075496

The books in this collection are devoted to challenging and open problems at the forefront of modern science and scholarship, including related philosophical debates. In contrast to typical research monographs, however, they strive to present their topics in a manner accessible also to scientifically literate non-specialists wishing to gain insight into the deeper implications and fascinating questions involved. Taken as a whole, the series reflects the need for a fundamental and interdisciplinary approach to modern science and research. Furthermore, it is intended to encourage active academics in all fields to ponder over important and perhaps controversial issues beyond their own speciality. Extending from quantum physics and relativity to entropy, consciousness, language and complex systems—the Frontiers Collection will inspire readers to push back the frontiers of their own knowledge.

More information about this series at http://www.springer.com/series/5342

Anthony Aguirre · Zeeya Merali · David Sloan
Editors

Undecidability, Uncomputability, and Unpredictability

 Springer

Editors
Anthony Aguirre
Physics Department
UC Santa Cruz
Santa Cruz, CA, USA

Zeeya Merali
Foundational Questions Institute
Decatur, GA, USA

David Sloan
Physics Department
Lancaster University
Lancaster, UK

ISSN 1612-3018   ISSN 2197-6619 (electronic)
THE FRONTIERS COLLECTION
ISBN 978-3-030-70353-0   ISBN 978-3-030-70354-7 (eBook)
https://doi.org/10.1007/978-3-030-70354-7

This Springer imprint is published by the registered company Springer Nature Switzerland AG
The registered company address is: Gewerbestrasse 11, 6330 Cham, Switzerland

Preface

This book is a collaborative project between Springer and The Foundational Questions Institute (FQXi). In keeping with both the tradition of Springer's Frontiers Collection and the mission of FQXi, it provides stimulating insights into a frontier area of science, while remaining accessible enough to benefit a non-specialist audience.

FQXi is an independent, non-profit organization that was founded in 2006. It aims to catalyze, support, and disseminate research on questions at the foundations of physics and cosmology.

The central aim of FQXi is to fund and inspire research and innovation that is integral to a deep understanding of reality, but which may not be readily supported by conventional funding sources. Historically, physics and cosmology have offered a scientific framework for comprehending the core of reality. Many giants of modern science—such as Einstein, Bohr, Schrödinger, and Heisenberg—were also passionately concerned with, and inspired by, deep philosophical nuances of the novel notions of reality they were exploring. Yet, such questions are often overlooked by traditional funding agencies.

Often, grant-making and research organizations institutionalize a pragmatic approach, primarily funding incremental investigations that use known methods and familiar conceptual frameworks, rather than the uncertain and often interdisciplinary methods required to develop and comprehend prospective revolutions in physics and cosmology. As a result, even eminent scientists can struggle to secure funding for some of the questions they find most engaging, while younger thinkers find little support, freedom, or career possibilities unless they hew to such strictures.

FQXi views foundational questions not as pointless speculation or misguided effort, but as critical and essential inquiry of relevance to us all. The Institute is dedicated to redressing these shortcomings by creating a vibrant, worldwide community of scientists, top thinkers, and outreach specialists who tackle deep questions in physics, cosmology, and related fields. FQXi is also committed to engaging with the public and communicating the implications of this foundational research for the growth of human understanding.

As part of this endeavor, FQXi organizes an annual essay contest, which is open to everyone, from professional researchers to members of the public. These contests are designed to focus minds and efforts on deep questions that could have a profound impact across multiple disciplines. The contest is judged by an expert panel and up to 20 prizes are awarded. Each year, the contest features well over a hundred entries, stimulating ongoing online discussion for many months even after the closure of the contest.

We are delighted to share this collection, inspired by the 2019–2020 contest, "Undecidability, Uncomputability, and Unpredictability." In line with our desire to bring foundational questions to the widest possible audience, the entries, in their original form, were written in a style that was suitable for the general public. In this book, which is aimed at an interdisciplinary scientific audience, the authors have been invited to expand upon their original essays and include technical details and discussion that may enhance their essays for a more professional readership, while remaining accessible to non-specialists in their field.

FQXi would like to thank our contest partners, the Fetzer Franklin Fund and The Peter and Patricia Gruber Foundation. The editors are indebted to FQXi's Scientific Director, Max Tegmark, and Managing Director, Kavita Rajanna, who were instrumental in the development of the contest. We are also grateful to Angela Lahee at Springer for her guidance and support in driving this project forward.

Santa Cruz, USA   Anthony Aguirre
London, UK   Zeeya Merali
Lancaster, UK   David Sloan
                              Foundational Questions Institute, www.fqxi.org, 2021

Contents

1   **Introduction** 1
Anthony Aguirre, Zeeya Merali, and David Sloan

2   **Undecidability and Unpredictability: Not Limitations,
but Triumphs of Science** 5
Markus P. Müller

3   **Indeterminism and Undecidability** 17
Klaas Landsman

4   **Unpredictability and Randomness** 47
Rade Vuckovac

5   **Indeterminism, Causality and Information: Has Physics Ever
Been Deterministic?** 63
Flavio Del Santo

6   **Undecidability, Fractal Geometry and the Unity of Physics** 81
T. N. Palmer

7   **A Gödelian Hunch from Quantum Theory** 97
Hippolyte Dourdent

8   **Epistemic Horizons: This Sentence Is $\frac{1}{\sqrt{2}}(|\text{true}\rangle + |\text{false}\rangle)$** 115
Jochen Szangolies

9   Why Is the Universe Comprehensible? 135
    Ian T. Durham

10 Noisy Deductive Reasoning: How Humans Construct Math,
    and How Math Constructs Universes 147
    David H. Wolpert and David Kinney

11 Computational Complexity as Anthropic Principle: A Fable 169
    Rick Searle

Appendix: List of Winners 179

Chapter 1
Introduction

Anthony Aguirre, Zeeya Merali, and David Sloan

> When I look at Gödel's proof of his undecidability theorem…The proof is a soaring piece
> of architecture, as unique and as lovely as Chartres Cathedral…It destroyed Hilbert's dream
> of reducing all mathematics to a few equations, and replaced it with a greater dream of
> mathematics as an endlessly growing realm of ideas…Every formalization of mathematics
> raises questions that reach beyond the limits of the formalism into unexplored territory.
>
> Freeman Dyson (2008).

For a brief time in history, it was possible to imagine that an advanced intellect could, given enough time and resources, in principle, understand how to mathematically prove everything that was true. Nineteenth-century polymath Pierre-Simon Laplace famously posited a demon that could discern what mathematics corresponds to physical laws, and use those laws to predict anything that happens before it happens, given sufficient information about all the particles in the universe.

But that time, in which scientists and philosophers envisaged a clockwork universe, has passed, following developments in mathematics, physics, and computer science. In 1931, German-Austrian mathematician Kurt Gödel published his undecidability results, in the form of his incompleteness theorems. These put paid to German mathematician David Hilbert's program to find a complete and consistent set of axioms for all of mathematics. Soon after, in 1937, English mathematician and computer scientist Alan Turing presented a proof of the conjecture that some decision problems are undecidable—that is, there is no single algorithm or computation that can infallibly give the right answer to some purely mathematical yes-no questions.

The world of physics was concurrently being revolutionised by the development of quantum theory, which suggested that nature is fundamentally indeterministic. The

A. Aguirre
Physics Department, UC Santa Cruz, Santa Cruz, CA, USA

Z. Merali (✉)
Foundational Questions Institute, Decatur, GA, USA

D. Sloan
Physics Department, Lancaster University, Lancaster, UK

outcomes of quantum experiments can never be predicted with certainty; instead, they appear to be set at random upon measurement. Meanwhile chaos theory stated that even complicated classical (non-quantum) processes would be plagued by unpredictability, in systems whose dynamics are highly sensitive to their initial conditions. Thus the twentieth century showed that there are rigorous arguments limiting what we can prove, compute, and predict.

In the intervening years, some connections between these results have come to light; however, many remain obscure, and their implications are unclear. Are there, for example, real consequences for physics of undecidability and non-computability? Are there implications for our understanding of the relations between agency, intelligence, mind, and the physical world?

These are some of the questions that were addressed by participants in FQXi's 2019–2020 essay contest, "Undecidability, Uncomputability, and Unpredictability." The contest drew over 200 entries from countries around the globe, including a diverse mix of entrants—from research physicists, philosophers, and computer scientists, to high-school students and interested non-academics. This volume brings together all 10 prize-winning entries.

Our joint first-prize winners were Markus Müller and Klaas Landsman. In chapter 2, "Undecidability and Unpredictability: Not Limitations but Triumphs of Science", Müller champions fundamental undecidability and unpredictability, arguing that for too long scientists have lamented them as barriers to fully understanding reality. He suggests that it is philosophically naïve to think that the answers to all questions are 'things' that exist ontologically, albeit inaccessible to science. Instead, Müller asserts, one should regard 'real patterns' as fundamental; such a shift implies that scientists can still strive to know all that there is to know.

In chapter 3, "Indeterminism and Undecidability", Landsman examines a famous theorem in the foundations of quantum physics by Irish physicist John Bell, and subsequent experiments based on it, that suggest that quantum theory is inherently indeterministic. The traditional reading of Bell's theorem leaves two loopholes that still allow for a fundamental deterministic framework to underly quantum theory: A deterministic theory can exist if it is 'non-local' (allowing signals to travel at faster-than-light speeds) or, alternatively, if measurement settings in experiments are somehow constrained by unknown external influences and thus cannot be said to be freely chosen. However, in his essay, Landsman claims that a precise analysis of undecidability and randomness in quantum theory closes off these two possibilities for determinism.

Many of our winners were similarly fascinated by the question of whether physical reality is truly indeterministic—a feature associated with the random and unpredictable outcomes of quantum experiments. Rade Vuckovac notes that such randomness seems to stand in opposition to determinism but, in chapter 4, "Unpredictability and Randomness", he argues that randomness could in fact be a consequence of determinism. Flavio Del Santo also takes an unconventional approach: Indeterministic quantum theory is often presented in contrast with deterministic classical physics but, in chapter 5, "Indeterminism, Causality and Information: Has Physics Ever

Been Deterministic?", Del Santo questions whether classical physics may itself be indeterministic, which in turn may have consequences for notions of causality.

In chapter 6, "Undecidability, Fractal Geometry and the Unity of Physics", Tim Palmer develops a causal deterministic description of quantum theory. His analysis invokes an uncomputable class of geometric model, in an effort to combine quantum theory, the general theory of relativity, and chaos theory, in one unified framework. Meanwhile, in chapter 7, "A Gödelian Hunch from Quantum Theory", Hippolyte Dourdent looks to Gödel's incompleteness theorems for inspiration in attempting to explain the source of numerous quantum paradoxes—with implications for the emergence of time. Dourdent builds, in part, on work by Jochen Szangolies, whose own essay discussing the notion of an 'epistemic horizon'—a bound on the amount of knowledge one can gain about a system—is presented in chapter 8, "Epistemic Horizons: This Sentence Is $1/\sqrt{2}(|\text{True}> + |\text{False}>)$", where it is applied to Bell's Theorem.

Other winners also specifically considered the issue of what can be known, and how it can be known. The question of how subjective observers can comprehend aspects of the universe and exploit that knowledge for gain is addressed by Ian Durham, in chapter 9, "Why Is the Universe Comprehensible?". David Wolpert and David Kinney use a computational model, in chapter 10, "Noisy Deductive Reasoning: How Humans Construct Math, and How Math Constructs Universes", to explain mathematical reasoning and how it provides a handle on the world. Finally, in chapter 11, "Computational Complexity as Anthropic Principle: A Fable", Rick Searle deems a Laplace demon untenable in the face of modern developments in physics and computational complexity, but argues that such constraints may help, rather than hinder, scientists' attempts to discover deeper truths about reality.

Given the contest topic, it is perhaps unsurprising that this compilation is dominated by researchers specialising in quantum foundations and computational complexity. Nonetheless it is gratifying that the question stimulated new insights not only into the nature of quantum reality, but also into how we construct knowledge and comprehend the universe, and potentially revealed new avenues for unifying quantum physics with general relativity. It thus seems that the optimistic outlook promoted by a number of our winners is correct: Undecidability, uncomputability, and unpredictability should not be bemoaned for limiting our understanding; rather these features open new windows on the fundamental nature of reality.

Reference

F. Dyson, *The Scientist as Rebel* (NRYB Collections: 2008)

Chapter 2
Undecidability and Unpredictability: Not Limitations, but Triumphs of Science

Markus P. Müller

Abstract It is a widespread belief that results like Gödel's incompleteness theorems or the intrinsic randomness of quantum mechanics represent fundamental limitations to humanity's strive for scientific knowledge. As the argument goes, there are truths that we can never uncover with our scientific methods, hence we should be humble and acknowledge a reality beyond our scientific grasp. Here, I argue that this view is wrong. It originates in a naive form of metaphysics that sees the physical and Platonic worlds as a collection of things with definite properties such that all answers to all possible questions exist ontologically somehow, but are epistemically inaccessible. This view is not only a priori philosophically questionable, but also at odds with modern physics. Hence, I argue to replace this perspective by a worldview in which a structural notion of *'real patterns'*, not *'things'* are regarded as fundamental. Instead of a limitation of what we can know, undecidability and unpredictability then become mere statements of *undifferentiation of structure*. This gives us a notion of realism that is better informed by modern physics, and an optimistic outlook on what we can achieve: we can know what there is to know, despite the apparent barriers of undecidability results.

2.1 The Pessimistic View

The early 20th century has given us insights into mathematics, physics, and computer science that seemed to shatter our hope for unlimited progress of scientific knowledge. In 1931, Gödel published his famous incompleteness theorems [1], implying that every sufficiently complex consistent axiomatic system contains statements that are true but unprovable within the system. An information-theoretic version of this

M. P. Müller (✉)
Institute for Quantum Optics and Quantum Information, Austrian Academy of Sciences, Boltzmanngasse 3, 1090 Vienna, Austria
e-mail: Markus.Mueller@oeaw.ac.at

Perimeter Institute for Theoretical Physics, 31 Caroline Street North, Waterloo, ON N2L 2Y5, Canada insight is Turing's proof of the unsolvability of the halting problem [2]: there is no algorithm that could, in all instances and in finite time, decide whether another specified computation will eventually halt or run indefinitely. At about the same time, the discovery of quantum physics has led us to the insight that Nature seems to be intrinsically random: even with maximal knowledge of the current state of the world, it is impossible to predict future events with certainty [3].

At first glance, these insights seem to point at limitations of science, suggesting an attitude of humility and disappointment. Before these results, there was David Hilbert's program to reduce all of mathematics to a finite, complete, provably consistent set of axioms [4]. And there was Pierre-Simon Laplace's famous declaration [5] of the possibility of a "demon", able to predict all of the future with certainty given sufficient physical data. The new insights were in stark contrast to these declarations, showing that the hopes of Hilbert, Laplace and others were misplaced. Does this mean that mathematics and physics are, as scientific disciplines, intrinsically deficient in some sense?

At least this is the way that these theorems are often portrayed, both in popular and in academic accounts. Regarding Gödel's theorems, Wikipedia [6] claims: *"These results set a limit in principle to mathematics: not every mathematical theorem can be formally derived or disproved from the axioms of some area [...] of mathematics."* This proposed limitation of mathematics is often contrasted to an alleged omnipotence of the human mind, leading to a class of "anti-mechanist" arguments: *"There have been repeated attempts to apply Gödel's theorems to demonstrate that the powers of the human mind outrun any mechanism or formal system"* [7]. Philosopher John R. Lucas [8] claims that *"given any machine which is consistent and capable of doing simple arithmetic, there is a formula it is incapable of producing as being true [...] but which we can see to be true"* (cited in [7]).

If, on the other hand, one gives up on the idea that the human mind is in some specific sense more powerful than any mechanism, then it may be tempting to read Gödel's theorem as a fundamental epistemic restriction for humanity. This view is vividly expressed, for example, by Driessen and Suarez [9]:

"In this book, recent mathematical theorems are discussed, which show that man never will reach complete mathematical knowledge. Also experimental evidence is presented that physical reality will always remain partially veiled to man, inaccessible to his control. It is intended to provide, in the various contributions, the pieces of a puzzle which restore the possibility of a natural, intellectual access to the existence of an omniscient and omnipotent being."

As the quotation indicates, there is a widespread view of quantum physics which regards its statistical character as a symptom of incompleteness. This view is defended, for example, by proponents of de Broglie–Bohm theory [10], a nonlocal hidden-variable interpretation of quantum mechanics. According to this view, there exists a deterministic underlying reality, and particles have well-defined positions at any time. However, the predictions of quantum mechanics are probabilistic due to fundamental uncontrollable disturbances. It is this unavoidable lack of experimental control that is ultimately responsible for the apparent randomness of measurement outcomes. According to this view, quantum theory's statistical character is thus most naturally interpreted as manifesting some fundamental epistemic restriction.

So has science, in the problems described above, hit an impenetrable barrier? Let us gain some intuition by looking at a problem where humanity seemed to hit a barrier for about two thousand years, before the problem became finally dissolved.

2.2 On Axiomatic Theories and Structural Differentiation

Around 300 BC, in his treatise *Elements*, Euclid formulated a set of axioms and postulates that were supposed to capture the essence of geometry [11]. One of these principles seemed less self-evident than the others and was hence standing out: Euclid's fifth postulate, the *parallel postulate*. Could this postulate perhaps be proven as a consequence of the others? This hope was the source of a twenty-centuries-long search for a proof, for a logical deduction of the uniqueness of any parallel line through any given external point from simpler assumptions.

In the 19th century, this search finally came to an end. The discovery of nonstandard geometry showed that the parallel postulate cannot logically follow from the others. When it became gradually clear that (what we call today) elliptical and hyperbolic geometries are consistent theories—and these theories satisfy all other principles *except* for the fifth postulate—the parallel postulate changed its status from an apparent necessity to a choice to be made.

Denote by T the totality of all of Euclid's axioms and postulates (or rather, of Hilbert's more rigorous reformulation [12]) *except* for the parallel postulate—we can see it as a formal system, or *theory*. But what is this theory T about? It describes geometric objects—points, lines, circles, and more—and their relations. It allows us to prove many interesting statements about these objects (such as: an exterior angle of a triangle is larger than either of the remote angles), but it leaves some questions undecided that seem natural to ask (is the sum of the angles in a triangle equal to 180°?). This theory is sometimes called *absolute geometry* (see [13] for details).

If we add the parallel postulate to T, we obtain another theory: T_1, familiar *Euclidean geometry*. On the other hand, we can add suitable modifications of the parallel postulate to T, and obtain T_2 and T_3: *hyperbolic* and *elliptic geometry*.

When we talk about a theory in this way, we mean a systems of axioms equipped with formal rules to generate theorems, formulated in some language. However, we typically have a mental picture of *the things that the theory talks about*—the "meaning" of the language. Euclidean geometry T_1, for instance, is typically not envisioned as an abstract language game, but vividly depicted as talking about *geometric structure*: geometric objects embedded in a plane. The standard mathematical description of this idea is to say that a theory can have a *model* [14]: a set (say, the two-dimensional plane) equipped with distinguished elements (such as points, lines, circles), functions, and relations (such as incidence or congruence) such that the theorems of the theory are true when understood as talking about these elements.

For what follows, we can take a somewhat different perspective which is implicitly shared by many practicing mathematicians, but rarely explicitly communicated. Let us stipulate the following informal definition.

> **Definition.** A *structure S* is whatever is described by a consistent theory T.

For example, the structure S_1 described by theory T_1 corresponds to the objects and relations of Euclidean geometry—points, lines, their incidences and congruences, and several others. More formally, we can define a structure S as the collection of all models of its theory T.

Figure 2.1 gives a sketch of the geometric structures mentioned above. Euclidean, hyperbolic and elliptic geometries S_1, S_2 and S_3 are *more differentiated structures* than absolute geometry S: they have all the properties of S, plus some additional properties. On the other hand, we can regard S as the collection of all its differentiations $S_1 \cup S_2 \cup S_3$, because every model for T_1, T_2 or T_3 is also a model for T.

> **Definition.** Structure S' is *more differentiated* than structure S if its corresponding theory T' is an extension of the theory T—that is, if T' contains the same formal rules and axioms as T, plus additional axioms. This implies that all models for T' are also models for T.
>
> Consequently, a structure is equal to the collection of all its differentiations.

The theory T describing absolute geometry is incomplete. That is, the question Q *"is there more than one line through any given point parallel to another line?"* is undecidable within T: neither the affirmation nor the negation of this question can be proven in T. However, we can still regard the corresponding structure S as a perfectly valid "thing" in some sense. It is simply the case that some questions (like Q) that we may ask about this "thing" *don't have answers*. An answer exists for differentiations S_1, S_2 and S_3 of S (no, yes, and no), but no answer to this question exists for S. And this is how it *must* be: since S is the collection of all its differentiations, neither affirmation nor negation[1] of Q can be true for S. In this sense, undecidability of Q in theory T simply refers to the fact that the corresponding structure S is *undifferentiated* with respect to Q. It is a bit like stem cells of the human embryo, which are undifferentiated in the sense that the question "what type of cells will these become?" does not (yet) have an answer.

But should we really regard S as a valid "structure" if it has these "holes" in its catalogue of properties? Isn't S a defective thing, since the corresponding theory T is defective, i.e. incomplete?

---

[1] Note that this does not violate the law of the excluded middle. The statement $A \vee \neg A$ (for A the answer to Q) is still true for S, precisely because it holds for all its differentiations.

Fig. 2.1 Absolute geometry
and its differentiations

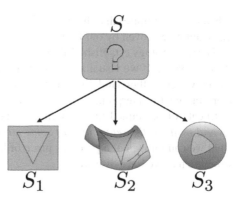

$$S$$

$$S_1 \qquad S_2 \qquad S_3$$

If we decided to throw out S on the basis of T's incompleteness, then we would quickly run out of interesting mathematical structures. This is precisely the content of *Gödel's first incompleteness theorem*:

> *"Any consistent theory T within which a certain amount of elementary arithmetic can be carried out is incomplete; i.e., there are statements of the language of T which can neither be proved nor disproved in T"* [7].

Thus (see also [7, Sect. 2.6]), we can obtain new theories T' and T'' by adding an unprovable statement *or its negation* as a new axiom to T. This means that the corresponding structure S has (at least) two inequivalent differentiations, S' and S''—similarly as absolute geometry has elliptic and hyperbolic (and Euclidean) geometries as differentiations. Calling a structure "interesting enough" if its theory T admits the necessary amount of arithmetic to be formalized for Gödel's theorem to apply, we arrive at the following consequence:

> **Theorem.** Every interesting enough structure has several inequivalent differentiations.

Intuitively, and perhaps traditionally, we tend to think of the mathematical world as consisting of "mathematical objects" with a catalogue of statements that are ontologically either true or false. For example, we may believe that there is something called "the natural numbers", $\mathbb{N}$, a well-established "thing" (after all, formalized as a *set*) that somehow "sits there", waiting for our mathematical tools to discover all of its properties and to prove all of its true theorems. Since mathematicians are only human, as this informal argument goes, all they can do is resort to theories T that try to capture the essence of $\mathbb{N}$ (such as the Peano axioms), and to use these theories to prove results about $\mathbb{N}$. According to this view, Gödel's first incompleteness theorem is bad news: it implies that there will always be statements about ℕ that are true, but that cannot be proven by our best theory.

The terminology established above allows us to take a different perspective on the Platonic world. If we visualize the mathematical world as consisting of structure in the above sense, then Gödel's first incompleteness theorem attains a quite different, more optimistic interpretation: proving the undecidability of a statement is not a certificate of principled human fallibility, but a deep insight into the *existence of several distinct differentiations of some structure*. It is not a fundamental limit to what we can know, but a precious piece of knowledge about a *non-property* of the structure that we have discovered.

2.3   The Physical World: Every Thing Must Go

Modern physics, as I will now argue, informs us that a similar move should be considered regarding our *physical* world. Consider the historical notion of the *luminiferous aether*. For several centuries, it was believed that light waves need a material medium for propagation, similarly as water waves. It was therefore natural to ask: *What are the properties of aether? How can we experimentally verify its existence and its properties?*

The historical course of events is well-known. After the exploration of light revealed more and more implausible properties of aether, the Michelson–Morley experiment and the subsequent development of Special Relativity has finally led the physics community to abandon this notion. This turn of events meant that the inability to answer the questions above (what are the properties of aether?) was not due to experimental limitations, but due to the fact that *the questions have no answers*. In other words: the questions were not solved, but *dissolved*.

The aether and its properties is by far not the only problem of physics with the final fate of dissolution. Consider the following question:

Did events A and B happen at the same time?

This is a very natural question with many highly relevant instances. For example, *did homo sapiens and homo erectus inhabit Southeast Asia at the same time?* Questions like this are immensely important for understanding our human ancestry [15]. *Did the suspect arrive at his hotel room at the same time that the victim was killed in the bar?* The answer may well determine whether the suspect is sent to prison.

Even though we can (hopefully) answer the question with enough effort in the cases just described, Special Relativity tells us that we cannot obtain an answer in all instances. But is this due to a limitation of our experimental abilities? No. According to Special Relativity, it is because *the above question doesn't have an answer*. It is only that in the cases of interest (such as the two just described), the question is implicitly asked relative to a predetermined frame of reference. Based on Newtonian mechanics, we thought that a general, absolute answer to this question *should* always exist, we have found that it doesn't.

The dissolution of questions tends to provoke considerable resistance. This is even true for well-established insights like the relativity of simultaneity, as the following quotation by philosopher and logician Peter Geach ([16], cited in [17]) demonstrates: *"[...] 'at the same time' belongs not to a special science but to logic. Our practical grasp of this logic is not to be called into question on account of recondite physics [...] A physicist who casts doubt upon it is sawing off the branch he sits upon".*

If such well-established instances of dissolution like the absence of absolute simultaneity provoke resistance, then it should not come as a surprise that such hesitation is particularly strong in the context of our second revolution of modern physics: quantum theory.

Quantum theory claims that there are questions that we may be interested in asking, but that we will never be able to answer, no matter what heavy artillery of physical methods we roll out. If we decide to prepare a quantum state in the superposition $|\psi\rangle = \frac{1}{\sqrt{2}}(|0\rangle + |1\rangle)$ and measure, we may be interested to know beforehand whether we will obtain outcome $|0\rangle$ or $|1\rangle$. But if quantum theory is correct, then this desired prediction is impossible. It is *as rock solid impossible* [18] as it is impossible to absolutely decide, in the regime of Special Relativity, whether two events A and B happened simultaneously.

Let us reformulate this observation of unpredictability. The question that turns out to be unanswerable is arguably best characterized as follows:

What is, at some given moment, the actual configuration *of the world?*

There are different ways to understand this question, depending on what we mean by an "actual configuration". In the foundations of quantum mechanics, this notion is often understood in a particular way: as a collection of values of ordinary variables that resemble what John Bell has called "beables" [19]. If such beables exist, and if they determine the outcomes of quantum measurements, then it is in principle impossible for us to get to know them all. These hypothetical "hidden variables" are not only epistemically inaccessible, but they also have properties that seem implausible. For example, by Bell's theorem, these hidden variables must be *nonlocal* in some sense; the way that they manifest themselves in measurements on entangled quantum states must necessarily involve superluminal signalling, but this signalling is miraculously washed out ("fine-tuned" [20]) so that it does not show up in the workings of our devices.

De Broglie–Bohm theory consists precisely of an attempt to answer the above question with an interpretation of the "actual configuration" of the world as just described. But also some proponents of ψ-epistemic interpretations [21, 22] *(what kind of hidden variables with what kind of causal structure give rise to quantum states as states of knowledge?)* or QBism [23] *(what kind of participatory real world gives rise to quantum states as rational states of belief?)* are strongly motivated by versions of this question. Arguing by analogy, we can characterize the situation in the structure terminology of Sect. 2.2 as follows. Quantum mechanics, as it is used in actual scientific practice, corresponds to some structure S. Bohmians claim that the world *actually* corresponds to a structure S' that is more differentiated than S, carrying unverifiable[2] answers to questions like "where is the electron exactly, right now?". Supporters of epistemic views in the sense above state as their goal to discover the correct differentiation S'.

The structure terminology suggests an obvious alternative: *perhaps the question simply doesn't* **have** *an answer*. The urge to claim that it does, but that we are unable to find it, is arguably motivated by a metaphysics of "things". Similarly as our example of the natural numbers $\mathbb{N}$ in Sect. 2.2, such a view of the world depicts the universe as a collection of objects, or as a thing in itself, that "sits there" in an infinitely differentiated form. Similarly as a material body *cannot not have a weight*, or a coin *cannot not show heads or tails*, we tend to take it as an analytic truth that the world *cannot not be in some configuration*. But if we see the world not as a thing, but as structure in some sense, then we may well accept the possibility that it corresponds more accurately to structure S than to any of its differentiations:

Interpretation. Modern physics has shown us that some apparent properties of the world are actually *non-properties*: they correspond to questions that do not *have* an answer.

While Gödel's results are *not* directly applicable to the physical world, they motivate a use of the *structure terminology* to interpret this phenomenon by analogy. Structure manifests itself by, and weaves together, "real patterns" [25] (such as correlations in measured data). Structure can be more or less differentiated. *Structural undifferentiation* means that there are questions that have no answers, or that there are less patterns than expected.

Apart from dissolving the question above, what else do we gain from a "structural" perspective? In some cases, the claim that a question doesn't have an answer can have surprising predictive power. As an example, consider *device-independent quantum cryptography* [26]. Two agents (Alice and Bob) perform local measurements on entangled quantum states. They use the random, correlated outcomes to generate a secret key. Could there be an eavesdropper (Eve) that spies on their message? If Alice's and Bob's statistics violates a Bell inequality, then the answer must be "no". Namely, if Eve is constrained by locality, and the setup violates *local realism*, then what remains is a phenomenal violation of realism: Eve cannot be correlated to any "elements of reality" in her past that would correspond to that secret key. In other words: *you cannot spy on something that doesn't exist*.

But the predictive power of this kind of reasoning seems to come at a price: aren't we giving up on *realism* here?

---

[2] Note that this characterization does not apply to Valentini's version of de Broglie–Bohm theory [24], which includes the possibility to have *quantum nonequilibrium systems* that make predictions which differ from standard quantum mechanics.

2.4 Ontic Structural Realism

In the context of quantum physics, the word "realism" is ambiguous and overloaded. The violation of a Bell inequality implies a violation of *local realism*, but these notions are defined in a very specific way. Interpretations of quantum theory that reject the violation of locality are often labelled as "anti-realist" [21]. But this includes interpretations of quantum mechanics that simply reject the mathematical framework on which the derivation is built in the first place (the ontological models framework [27]), even if they rest on a generally realist view of the physical world. A prominent example is given by QBism [23] that subscribes to a notion of "participatory realism".

In particular, to be a realist doesn't commit one to a metaphysics of "things"—perhaps quite on the contrary. This is the main point in a book by Ladyman et al. with the title of the previous section: Every Thing Must Go [17]. The authors argue precisely for a form of realism relying on a metaphysics of structure, not things—*ontic structural realism*.

The goals of Ladyman et al. are quite different from those of this essay—they are mostly motivated by the problems of standard scientific realism: *"[...] the history of successful novel prediction science is the most compelling evidence for some form of realism, but [...] the history of ontological discontinuity across theory change makes standard scientific realism indefensible."*

In particular, what is rejected is the 'doctrine of containment': *"On this doctrine, the world is a kind of container bearing objects that change location and properties over time. These objects cause things to happen by interacting directly with one another. [...] they themselves are containers in turn, and their properties and causal dispositions are to be explained by the properties and dispositions of the objects they contain (and which are often taken to comprise them entirely)."*

It is argued that this kind of view is in line with human intuition, but not so much with modern physics: *"we should not interpret science [...] as metaphysically committed to the existence of self-subsistent individuals. [...] We will later say that what exists are ('real') 'patterns'. [...] When we go on to deny that, strictly speaking, there are 'things', we will mean to deny that in the material world as represented by the currently accepted scientific structures, individual objects have any distinctive status."*

Such a version of realism is still able to account for the "no-miracles argument" [28]: that the best explanation of the success of science is that our best scientific theories are at least approximately true. In particular, it frees us from a problem famously described by Laudan [29]: if we insist on understanding "approximate truth" as the property that the central terms of a scientific theory (such as Dalton's atoms or Bohr's electrons) refer to actual entities in the world, then we have to regard previous physical theories (and thus perhaps also contemporary ones) as utterly unsuccessful. On the other hand, if we base realism on a structural ontology of "real patterns", then this problem is dissolved, and a form of stability across theory change is established.

In summary: applying structural terminology to our understanding of the physical world is not in conflict with realism, but on the contrary implied by mature versions of it.

2.5 Quantum-Optimistic Conclusions

What can we conclude if we accept the structural view put forward in this essay? First, interpreting undecidability as undifferentiation of structure arguably renders "anti-mechanist" views as expressed for example by Driessen and Suarez [9] implausible. *Non-existing* answers can neither be found by machines, nor by humans—nor by gods.

Regarding quantum theory, the structural perspective seems to bring us closer to views that are often unduly characterized as anti-realist: views in which quantum states are states of information about future measurement outcomes (or experiences), but not about some underlying reality [21]. But we do not have to stop here. Seeing the world as consisting of real patterns interwoven by structure, not as a "thing" or a collection of things, opens up the possibility to reconcile these epistemic interpretations with others that regard the quantum state as ontic. Namely, if quantum states are expressions of knowledge (or belief, or chance), and if the quantum state "is" the world (or is in the world), then why not accept the conjunction of both views?

> **Hypothesis.** The quantum world **is** probabilistic structure. In other words, it is not a "thing" or a collection of things, but it is the multitude of statistical patterns and their structural relations that any observer encounters in their data.

In [30], I have worked out a concrete version of this hypothesis in detail. It starts with the claim that the world *is nothing but* the determination of the chances of what happens to any observer next, and derives our usual picture of a "thing-like" objective world from it. Regardless of this specific approach, the main message is one of *optimism*: seeing unpredictability not as an expression of a fundamental epistemic restriction, but as structural undifferentiation admits new fruitful perspectives on the world, including ones that drop the false dichotomy of "ontic" and "epistemic" interpretations of the quantum state.

In this view, undecidability and unpredictability are not in themselves sufficient reasons to be pessimistic. But perhaps this is a dangerous perspective. As Steven Pinker [31] points out,

Pessimism has been equated with moral seriousness. Citing the popular naysayers, if you think knowledge can help solve problems, then you have a "blind faith" and a "quasi-religious belief" in the "outmoded superstition" and "false promise" of the "myth" of the "onward march" of "inevitable progress".

In the light of Gödel's theorems, the epitaph of David Hilbert's tombstone in Göttingen is sometimes regarded as a prototype of a false promise:
We must know. We will know.
Let me therefore conclude this essay with one further outmoded declaration of blind faith:
We can know what there is to know.

References

1. K. Gödel, Über formal unentscheidbare Sätze der Principia Mathematica und verwandter Systeme, I. Monatshefte für Mathematik und Physik **38**(1), 173–198 (1931)
2. A. Turing, On computable numbers, with an application to the Entscheidungsproblem. Proc. Lond. Math. Soc. Ser. 2 **42**, 230–265 (1937)
3. A. Einstein, B. Podolsky, N. Rosen, Can quantum-mechanical description of physical reality be considered complete? Phys. Rev. **47**(10), 777–780 (1935)
4. R. Zach, Hilbert's program, in *The Stanford Encyclopedia of Philosophy (Fall 2019 Edition)*, ed. by E.N. Zalta, https://plato.stanford.edu/archives/fall2019/entries/hilbert-program/
5. P.-S. Laplace, Essai philosophique sur les probabilités, introduction to Théorie Analytique des Probabilités (1820)
6. Wikipedia, Gödelscher Unvollständigkeitssatz, https://de.wikipedia.org/wiki/Gödelscher_Unvollständigkeitssatz. Accessed 1 Apr 2020
7. P. Raatikainen, Gödel's incompleteness theorems, in *The Stanford Encyclopedia of Philosophy (Summer 2020 Edition)*, ed. by E.N. Zalta, forthcoming, https://plato.stanford.edu/archives/sum2020/entries/goedel-incompleteness/
8. J.R. Lucas, Minds, machines, and Gödel. Philosophy **36**(137), 112–137 (1961)
9. A. Driessen, A. Suarez (eds.), *Mathematical Undecidability, Quantum Nonlocality and the Question of the Existence of God* (Springer, Dordrecht, 1997)
10. D. Bohm, A suggested interpretation of the quantum theory in terms of "hidden" variables. I. Phys. Rev. **85**(2), 166–179 (1952)
11. Euclid, *The Thirteen Books of Euclid's Elements*, translation and commentaries by Sir T.L. Heath (Dover Publications, New York, 1956)
12. D. Hilbert, *The Foundations of Geometry* (transl. L. Unger from the 10th German ed.), 2nd English ed., revised and enlarged by P. Bernays, Open Court, Peru, IL (1988)
13. M.J. Greenberg, Old and new results in the foundations of elementary plane Euclidean and non-Euclidean geometries. Am. Math. Mon. **117**, 198–219 (2010)
14. D. Marker, Introduction to model theory. Model Theory Algebra Geom. **39**, 15–35 (2000)
15. C.C. Swisher III, W.J. Rink, S.C. Antón, H.P. Schwarcz, G.H. Curtis, A. Suprijo Widiasmoro, Latest Homo erectus of Java: potential contemporaneity with Homo sapiens in Southeast Asia. Science **274**, 5294 (1996)
16. P. Geach, *Logic Matters* (Oxford University Press, Oxford, 1972)
17. J. Ladyman, D. Ross, D. Spurrett, J. Collier, *Every Thing Must Go - Metaphysics Naturalized* (Oxford University Press, Oxford, 2007)
18. R. Colbeck, R. Renner, No extension of quantum theory can have improved predictive power. Nat. Commun. **2**, 411 (2011)
19. J.S. Bell, The theory of local beables, *Speakable and Unspeakable in Quantum Mechanics*, 2nd edn. (Cambridge University Press, Cambridge, 2004)
20. C.J. Wood, R.W. Spekkens, The lesson of causal discovery algorithms for quantum correlations: causal explanations of Bell-inequality violations require fine-tuning. New J. Phys. **17**, 033002 (2015)

21. M.S. Leifer, Is the quantum state real? An extended review of ψ-ontology theorems. Quanta **3**(1), 67–155 (2014)
22. R.W. Spekkens, Evidence for the epistemic view of quantum states: a toy theory. Phys. Rev. A **75**, 032110 (2007)
23. C. Fuchs, On participatory realism, in *Information and Interaction: Eddington, Wheeler, and the Limits of Knowledge*, ed. by I.T. Durham, D. Rickles (The Frontiers Collection, Springer, Cham, 2017)
24. N.G. Underwood, A. Valentini, Quantum field theory of relic nonequilibrium systems. Phys. Rev. D **92**, 063531 (2015)
25. D.C. Dennett, Real patterns. J. Philos. **88**(1), 27–51 (1991)
26. J. Barrett, L. Hardy, A. Kent, No signaling and quantum key distribution. Phys. Rev. Lett. **95**, 010503 (2005)
27. N. Harrigan, R.W. Spekkens, Einstein, incompleteness, and the epistemic view of quantum states. Found. Phys. **40**, 125 (2010)
28. H. Putnam, Philosophy and our mental life, *Mind, Language, and Reality* (Cambridge University Press, New York, 1975), pp. 291–303
29. L. Laudan, A confutation of convergent realism. Philos. Sci. **48**(1), 19–49 (1981)
30. M.P. Müller, Law without law: from observer states to physics via algorithmic information theory. Quantum **4**, 301 (2020)
31. S. Pinker, *Enlightenment Now - The Case for Reason, Science, Humanism, and Progress* (Penguin Random House, New York, 2018)

Chapter 3
Indeterminism and Undecidability

Klaas Landsman

Dedicated to the memory of Michael Redhead (1929–2020).

Abstract The aim of this paper is to argue that the (alleged) indeterminism of quantum mechanics, claimed by adherents of the Copenhagen interpretation since Born [8], can be proved from Chaitin's follow-up to Gödel's (first) incompleteness theorem. In comparison, Bell's [2] theorem as well as the so-called free will theorem-originally due to Heywood and Redhead [44]-left two loopholes for deterministic hidden variable theories, namely giving up either *locality* (more precisely: local contextuality, as in Bohmian mechanics) or *free choice* (i.e. uncorrelated measurement settings, as in 't Hooft's cellular automaton interpretation of quantum mechanics). The main point is that Bell and others did not exploit the full empirical content of quantum mechanics, which consists of long series of outcomes of repeated measurements (idealized as infinite binary sequences): their arguments only used the long-run relative frequencies derived from such series, and hence merely asked hidden variable theories to reproduce single-case Born probabilities defined by certain entangled bipartite states. If we idealize binary outcome strings of a fair quantum coin flip as infinite sequences, quantum mechanics predicts that these typically (i.e. almost surely) have a property called *1-randomness* in logic, which is much stronger than uncomputability. This is the key to my claim, which is admittedly based on a stronger (yet compelling) notion of determinism than what is common in the literature on hidden variable theories.

K. Landsman (✉)
Department of Mathematics, Institute for Mathematics, Astrophysics, and Particle Physics (IMAPP), Radboud University, Nijmegen, The Netherlands
e-mail: landsman@math.ru.nl

© The Author(s), under exclusive license to Springer Nature Switzerland AG 2021
A. Aguirre et al. (eds.), *Undecidability, Uncomputability, and Unpredictability*,
The Frontiers Collection, https://doi.org/10.1007/978-3-030-70354-7_3

3.1 Introduction: Gödel and Bell

While *prima facie* totally unrelated, Gödel's theorem [37] in mathematical logic and Bell's theorem [2] in physics share a number of fairly unusual features (for *theorems*)[1]:

- Despite their very considerable technical and conceptual difficulty, both results are extremely famous and have caught the popular imagination like few others in science.
- Though welcome in principle-in their teens, many people including the author were intrigued by books with titles like *Gödel, Escher Bach: An Eternal Golden Braid* and *The Dancing Wu-Li Masters: An Overview of the New Physics*, both of which appeared in 1979-this imagination has fostered wild claims to the effect that Gödel proved that the mind cannot be a computer or even that God exists, whilst Bell allegedly showed that reality does not exist. Both theorems (apparently through rather different means) supposedly also supported the validity of Zen Buddhism.[2]
- However, even among professional mathematicians (logicians excepted) few would be able to correctly state the content of Gödel's theorem when asked on the spot, let alone provide a correct proof, and similarly for Bell's theorem among physicists.
- Nonetheless, many professionals will be aware of the general feeling that Gödel in some sense shattered the great mathematician Hilbert's dream of what the foundations of mathematics should look like, whilst there is similar consensus that Bell dealt a lethal blow to Einstein's physical world view-though ironically, Gödel worked in the spirit and formalism of Hilbert's proof theory, much as Bell largely *agreed* with Einstein's views about quantum mechanics and about physics in general.
- Both experts and amateurs seem to agree that Gödel's theorem and Bell's theorem penetrate the very core of the respective disciplines of mathematics and physics.

In this light, anyone interested in both of these disciplines will want to know what these results have to do with each other, especially since mathematics underwrites physics (or at least is its language).[3] At first sight this connection looks remote. Roughly speaking[4]:

---

[1] In fact there are *two* incompleteness theorems in logic due to Gödel (see footnotes 2 and 5) and *two* theorems on quantum mechanics due to Bell [12, 89], but for reasons to follow in this essay I am mainly interested in the first ones, of both authors, except for a few side remarks.

[2] See Franzén [35] for an excellent first introduction to Gödel's theorems, combined with a fair and detailed critique of its abuses, including overstatements by both amateurs and experts (a similar guide to the use and abuse of Bell's theorems remains to be written), and Smith [75] for a possible second go.

[3] Yanofsky [91] nicely discusses both theorems in the context of the limits of science and reason.

[4] Both reformulations are a bit anachronistic and purpose-made. See Gödel [37] and Bell [2]!

1. Gödel proved that any consistent mathematical theory (formalized as an axiomatic-deductive system in which proofs could in principle be carried out mechanically by a computer) that contains enough arithmetic is incomplete (in that arithmetic sentences φ exist for which neither φ nor its negation can be proved).

2. Bell showed that if a deterministic "hidden variable" theory underneath (and compatible with) quantum mechanics exists, then this theory cannot be local (in the sense that the hidden state, if known, could be used for superluminal signaling).

Both were triggered by a specific historical context. Gödel [37] reflected on the recently developed formalizations of mathematics, of which he specifically mentions the *Principia Mathematica* of Russell and Whitehead and the axioms for set theory proposed earlier by Zermelo, Fraenkel, and von Neumann. Though relegated to a footnote, the shadow of *Hilbert's program*, aimed to prove the consistency of mathematics (ultimately based on Cantor's set theory) using absolutely reliable, "finitist" means, clearly loomed large, too.[5]

Bell, on the other hand, tried to understand if the *de Broglie–Bohm pilot wave theory*, which was meant to be a deterministic theory of particle motion reproducing all predictions of quantum mechanics, *necessarily* had to be non-local: Bell's answer, then, was "yes."[6]

In turn, the circumstances in which Gödel and Bell operated had a long pedigree in the quest for *certainty in mathematics* and for *determinism in physics*, respectively.[7] The former had even been challenged at least three times[8]: first, by the transition from Euclid's mathematics to Newton's; second, by the set-theoretic paradoxes discovered around 1900 by Russell and others (which ultimately resulted from attempts to make Newton's calculus rigorous by grounding it in analysis, and in turn founding analysis in the real numbers and hence in set theory), and third, by Brouwer's challenge to "classical" mathematics, which he tried to replace by "intuitionistic" mathematics (both Hilbert and Gödel were influenced by Brouwer, though *contrecoeur*: neither shared his overall philosophy of mathematics).

---

[5] Gödel's *second* incompleteness theorem shows that one example of φ is the (coded) statement that the consistency of the theory can be proved within the theory. This is often taken to refute Hilbert's program, but even among experts it seems controversial if it really does so. For Hilbert's program and its role in Gödel's theorems see e.g. Zach [92], Tait [83], Sieg [74], and Tapp [84].

[6] Greenstein [39] is a popular book on the history and interpretation of Bell's work. Scholarly analyses include Redhead [68], Butterfield [14], Werner and Wolf [88], and the papers cited in footnote 49.

[7] Some vocal researchers calim that Bell and Einstein were primarily interested in locality and realism, determinism being a secondary (or no) issue, but the historical record is ambiguous; more generally, over 10,000 papers about Bell's theorems show that Bell can be interpreted in almost equally many ways. But this controversy is a moot point: whatever his own (or Einstein's) intentions, Bell's [2] theorem puts constraints on possible deterministic underpinnings of quantum mechanics, and that is how I take it.

[8] For an overall survey of this theme see Kline [47].

In physics (and more generally), what Hacking ([42], Chap. 2) calls the *doctrine of necessity*, which thus far-barring a few exceptions-had pervaded European thought, began to erode in the 19th century, culminating in the invention of quantum mechanics between 1900–1930 and notably in its probability interpretation as expressed by Born [8]:

> Thus Schrödinger's quantum mechanics gives a very definite answer to the question of the outcome of a collision; however, this does not involve any causal relationship. One obtains *no* answer to the question "what is the state after the collision," but only to the question "how probable is a specific outcome of the collision". (...) This raises the entire problem of determinism. From the standpoint of our quantum mechanics, there is no quantity that could causally establish the outcome of a collision in each individual case; however, so far we are not aware of any experimental clue to the effect that there are internal properties of atoms that enforce some particular outcome. Should we hope to discover such properties that determine individual outcomes later (perhaps phases of the internal atomic motions)? (...) I myself tend to relinquish determinism in the atomic world. ([8], p. 866, translation by the present author)

In a letter to Born dated December 4, 1926, Einstein's famously replied that 'God does not play dice' ('Jedenfalls bin ich überzeugt, daß *der* nicht würfelt'). Within ten years Einstein saw a link with locality,[9] and Bell [2] and later papers followed up on this.

3.2 Randomness and Its Unprovability

This precise history has a major impact on my argument, since it shows that right from the beginning the kind of randomness that Born (probably preceded by Pauli and followed by Bohr, Heisenberg, Jordan, Dirac, von Neumann, and most of the other pioneers of quantum mechanics except Einstein, de Broglie, and Schrödinger) argued for as being produced by quantum mechanics, was antipodal to *determinism*.[10] Thus randomness in quantum mechanics was identified with *indeterminism*, and hence attempts (like the de Broglie–Bohm pilot wave theory) to undermine the "Copenhagen" claim of randomness looked for *deterministic* (and arguably *realistic*) theories underneath quantum mechanics.

Although "undecidability" may sound a bit like "indeterminism", the analogy between the quests for certainty in mathematics and for determinism in physics (and their alleged undermining by Gödel's and Bell's theorems, respectively) may sound rather superficial. To find common ground more effort is needed to bringing these theorems together.[11]

---

[9] This phase in the history of quantum mechanics is described by Mehra and Rechenberg [57].

[10] See Landsman [53] for the view that randomness is a family resemblance (in that it lacks a meaning common to all its applications) with the special feature that its various uses are always defined antipodally.

[11] Also cf. e.g. Breuer [9], Calude [16], Svozil, Calude and Stay [79], and Szangolies [80].

First, some of its "romantic" aspects have to be removed from Gödel's theorem, notably its reliance on self-reference, although admittedly this *was* the key to both Gödel's original example of an undecidable sentence φ (which in a cryptic way expresses its own unprovability) and his proof, in which an axiomatic theory that includes arithmetic is arithmetized through a numerical encoding scheme so as to be able to "talk about itself". Though later proofs of Gödel's theorem also use numerical encodings of mathematical expressions (such as symbols, sentences, proofs, and computer programs), this is done in order to make recursion theory (initially a theory of functions $f : \mathbb{N} \to \mathbb{N}$) available to a wider context, rather than to exploit self-reference. Each computably enumerable but uncomputable subset $E \subset \mathbb{N}$ leads to undecidable statements (very rarely in mainstream mathematics),[12] namely those for which the sentence $n \notin E$ is true but unprovable. *Chaitin's (first) incompleteness theorem* (Theorem B.1 in Appendix B), which will play an important role in my reasoning, is an example of this. To understand this theorem and its background we return to the history of 20th century mathematics and physics.

Hilbert influenced this history in many ways,[13] of which the sixth problem on his famous list of 23 mathematical problems from 1900 is particularly relevant here: this problem concerns the '*Mathematical Treatment of the Axioms of Physics, especially the theory of probabilities and mechanics*' [45]. This problem influenced our topic in two initially independent ways, which now come together. First, the problem inspired von Neumann [62] to develop his mathematical axiomatization of quantum mechanics, which still forms the basis of all mathematically rigorous work on this theory. In particular, he initiated the literature on hidden variable theories (see Sect. 3.3). Second, it led both von Mises and Kolmogorov to their ideas on the mathematical foundations of probability and randomness, initially in opposite ways: whereas von Mises [58] tried (unsuccessfully) to first axiomatize random sequences of numbers and then extract probability from these as asymptotic relative frequency, Kolmogorov [49] successfully axiomatized probability first and then (unsuccessfully) sought to extract some notion of randomness from this.

The basic problem (already known to Laplace and perhaps even earlier probabilists) was that, in a 50-50 Bernoulli process for simplicity, an apparently "random" string like

$$\sigma = 0011010101110100101000110110111 \tag{3.2.1}$$

is as probable as a "deterministic" string like

$$\sigma = 1111111111111111111111111111111. \tag{3.2.2}$$

---

[12] A subset $E \subset \mathbb{N}$ is *computably enumerable* (c.e.) if it is the image of a computable function $f : \mathbb{N} \to \mathbb{N}$, and *computable* if its characteristic function 1_E is computable, which is true iff both E and $\mathbb{N} \backslash E$ are c.e.

[13] This is true for physics almost as much as it is (more famously) for mathematics, since Hilbert played a major role in the mathematization of the two great theories of twentieth century physics, i.e. general relativity [22, 69] and quantum mechanics [54, 67]).

In other words, their probabilities say little or nothing about the "randomness" of individual outcomes. Imposing statistical properties helps but is not enough to guarantee randomness. It is slightly easier to explain this in base 10, to which I therefore switch for a moment. If we call a sequence x *Borel normal* if each possible string σ in x has relative frequency $10^{-|\sigma|}$, where $|\sigma|$ is the length of σ (so that each digit $0, \ldots, 9$ occurs 10% of the time, each block 00 to 99 occurs 1% of the time, etc.), then *Champernowne's number*

$$01234567891011121314151617181920212223242526292829 30 \ldots$$

can be shown to be Borel normal. The decimal expansion of π is conjectured to be Borel normal, too (and has been empirically verified to be so in billions of decimals), but these numbers are hardly random: they are computable, which is one version of "deterministic".

Any sound definition of randomness (for binary strings or sequences) has to navigate between Scylla and Charybdis: if the definition is too weak (such as Borel normality), counterexamples will undermine it (such as Champernowne's number), but if it is too strong (such as being lawless, like Brouwer's choice sequences), it will not hold almost surely in a 50-50 Bernoulli process [60]. As an example of such sound navigation, Solomonoff, Kolmogorov, Martin-Löf, Chaitin, Levin, Solovay, Schnorr, and others developed the *algorithmic theory of randomness* [56]. The basic idea is that a string or sequence is random iff its shortest description is the sequence itself, but the notion of a description has to made precise to avoid *Berry's paradox*:

The Berry number is the smallest positive integer that cannot be described in less than eighteen words.

The paradox, then, is that on the one hand this number must exist, since only finitely many integers can be described in less than eighteen words and hence the set of such numbers must have a lower bound, while on the other hand Berry's number cannot exists by its own definition.[14] In the case at hand, the notion of a *description* is sharpened by asking it to be *computable*, so that, roughly speaking (see Appendix B for technical details), we call a (finite) binary string σ *(Kolmogorov) random* if the length of the shortest computer program generating σ is at least as long as σ itself, and call an (infinite) binary sequence x *(Levin–Chaitin) random* or *1-random* if its (sufficiently long) finite truncations are Kolmogorov random. At last, for finite strings σ Chaitin's *(first) incompleteness theorem* states that although countably many strings σ *are* random, this can be *proved* only for finitely many of these, whereas for infinite sequences x his (second) incompleteness theorem says that if such a sequence is random, only finitely many of its digits can be computed (see Theorems B.1 and B.4 for precise statements). Thus *randomness is elusive*.

---

[14] This is one of innumerable paradoxes of natural language, which leads to an incompleteness theorem once the notion of a description has been appropriately formalized in mathematics, much as Gödel's first incompleteness theorem turns the liar's paradox into a theorem.

3.3 Rethinking Bell's Theorem

In order to locate Bell's [2] theorem in the literature on quantum mechanics and (in)determinism, I recall that Hilbert's sixth problem inspired both the work of von Mises and Kolmogorov that eventually gave rise to the algorithmic theory of randomness, *and* (Hilbert's postdoc) von Neumann's work on the mathematical foundations of quantum mechanics, culminating in his book [62]. One of his results was that there can be no nonzero function $\lambda : H_n(\mathbb{C}) \to \mathbb{R}$ (where $H_n(\mathbb{C})$ is the space of hermitian $n \times n$ matrices, seen as the observables of a quantum-mechanical n-level system) that is:

1. *dispersion-free* (i.e. $\lambda(a^2) = \lambda(a)^2$ for each $a \in H_n(\mathbb{C})$);
2. *linear* (i.e. $\lambda(sa + tb) = s\lambda(a) + t\lambda(b)$ for all $s, t \in \mathbb{R}$ and $a, b \in H_n(\mathbb{C})$).

Unfortunately, von Neumann interpreted this correct, non-circular, and interesting result as a proof that quantum mechanics is complete in the sense that there can be no hidden variables in the sense of Born [8], i.e. 'properties that determine individual outcomes'. The reason this does not follow is twofold.[15] First, the proof relies on a tacit assumption that later came to be called *non-contextuality*, namely that the value $\lambda(a)$ of some observable a *only depends on* a, whereas measurement ideology à la Bohr [7] suggests that it may depend on a *measurement context*, formalized as a further set of observables commuting with a (unless a is maximal such a set is far from unique).[16] Second, though natural, the linearity assumption is very strong and excludes even eigenvalues of a.

This second point was remedied by the *Kochen–Specker theorem*,[17] who weakened von Neumann's linearity assumption to *linearity on commuting observables*, which at least incorporates eigenvalues and is even found so appealing that the Kochen–Specker is generally taken to exclude non-contextual hidden variable theories. See also Appendix C.

The final step in the series of attempts, initiated by von Neumann, to exclude hidden variables by showing that subject to reasonable assumptions the corresponding value attributions cannot exist even independently of any statistical considerations, is the so-called *free will theorem*.[18] In the wake of the renowned "EPR" paper [33] the setting has now become bipartite (i.e. Alice and Bob who are spacelike separated each perform experiments on a correlated state) and the *non-contextuality* assumption is weakened to *local contextuality*: the outcomes of Alice's measurements are indepen-

---

[15] See also Bub [13], Dieks [26], and forthcoming work by Chris Mitsch for balanced accounts.

[16] The idea of contextuality was first formulated by Grete Hermann [23].

[17] See Kochen and Specker [48]. Ironically, his followers attribute this theorem to Bell [3], although the result is just a technical sharpening of von Neumann's result they so vehemently ridicule. For a deep philosophical analysis of the Kochen–Specker theorem, as well as of Bell's theorems, see Redhead [68].

[18] See Appendix C. This theorem is originally due to Heywood and Redhead [44], with follow-ups by Stairs [76], Brown and Svetlichny [11], and Clifton [18], but it was named and made famous by Conway and Kochen [21], whose main contribution was an emphasis on free will ([52], Chap. 6).

dent of any choice of measurements Bob might perform, and *vice versa*.[19] Thus her value attributions $\lambda(a|context)$ may well be contextual, *as long as the observables commuting with the one she measures (i.e. a), which form a context to a, are local to her.*

A second line of research, which goes back at least to de Broglie [10], was influentially taken up by Bohm [6], and most recently includes 't Hooft [46], assumes the possibility of non-contextual value attributions and tries to make these compatible with the Born rule of quantum mechanics. Bell [2] was primarily concerned with such theories, asking himself if a deterministic theory like Bohm's was necessarily non-local.

In Bell's analysis, which takes place in the bipartite (EPR) setting, the quantum-mechanical probabilities are obtained by formally averaging over the set of hidden variables, i.e.,

$$P_\psi(F = x, G = y \mid A = a, B = b) = \int_\Lambda d\mu_\psi(\lambda)\, P_\lambda(F = x, G = y \mid A = a, B = b).$$
(3.3.1)

Here ψ is some (explicitly identified) quantum state of a correlated pair of (typically) 2-level quantum systems (which may be either optical, where the degree of freedom is helicity, or massive, where the degree of freedom is spin), F is an observable measured by Alice defined by her choice of setting a, likewise G for Bob defined by his setting b, with possible outcomes $x \in \underline{2} = \{0, 1\}$, likewise $y \in \underline{2}$ for Bob; the left-hand side is the Born probability for the outcome (x, y) if the correlated system has been prepared in the state ψ; the expression $P_\lambda(\cdots)$ on the right-hand side is the probability of the outcome (x, y) if the unknown hidden variable or state equals λ, and finally, μ_ψ is some probability measure on the space Λ of hidden states supposedly provided by the theory for each state ψ.

We now say that the hidden variable theory supplying the above quantities is:

- *deterministic* if the probabilities $P_\lambda(F = x, G = y \mid A = a, B = b)$ equal 0 or 1;
- *locally contextual* if the expression

$$P_\lambda(F = x \mid A = a, B = b) = \sum_{y=0,1} P_\lambda(F = x, G = y \mid A = a, B = b);$$
(3.3.2)

is independent of b, whilst the corresponding expression

$$P_\lambda(G = y \mid A = a, B = b) = \sum_{x=0,1} P_\lambda(F = x, G = y \mid A = a, B = b),$$
(3.3.3)

is independent of a. That is, the probabilities of Alice's outcomes are independent of Bob's settings, and *vice versa*. This locality property seems very reasonable and in fact it follows from special relativity, for if Bob chooses his settings just before his measurement, there is a frame of reference in which Alice measures

---

[19] Since Alice and Bob are spacelike separated their observables commute (Einstein locality).

before Bob has chosen his settings, and *vice versa*. In turn, this is equivalent to the property that even if she knew the value of λ, Alice could not signal to Bob, and *vice versa*.[20]

Bell proved that a hidden variable theory cannot satisfy (3.3.1) and be both deterministic and locally contextual (which explained why Bohm's theory had to be nonlocal). Making his *tacit* assumption that experimental settings can be "freely" chosen *explicit*, we obtain[21]:

Theorem 3.3.1 *The conjunction of the following properties is inconsistent:*

1. determinism;
2. quantum mechanics, *i.e. the Born rule for* $P_\psi(F = x, G = y \mid A = a, B = b)$;
3. local contextuality;
4. free choice, *i.e. (statistical) independence of the measurement settings a and b from each other and from the hidden variable* λ *(given the probability measure* μ_ψ*).*

3.4 Are Deterministic Hidden Variable Theories Deterministic?

Although the assumptions have a slightly different meaning, the free will theorem leads to the same result as Bell's theorem (see Appendix 3), so that the (no) hidden variable tradition initiated by von Neumann, which culminates in the former, coalesces with the (positive) hidden variable tradition going back to de Broglie, shown its place by the latter. Thusly there are the obvious four (minimal) ways out of the contradiction in Theorem 3.3.1:

- Copenhagen (i.e. mainstream) quantum mechanics rejects determinism;
- Valentini [86] rejects the Born rule and hence QM (see the end of Sect. 3.5 below);
- Bohmians reject local contextuality[22];
- 't Hooft [46] rejects free choice.

We focus on the last two options, so that determinism and quantum mechanics (i.e. the Born rule) are kept. In both cases the Born rule is recovered by averaging the hidden variable with respect to a probability measure μ_ψ on the space of hidden variables, given some (pure) quantum state ψ. The difference is that in Bohmian mechanics the total state (which consists of the hidden configuration plus the "pilot wave" ψ) determines the measurement outcomes *given the settings*, whereas in 't

---

[20] In quantum mechanics the left-hand side of (3.3.1) satisfies this locality condition for any state ψ.

[21] See [52], §6.5 for details, or Appendix C below for a summary.

[22] There is a subtle difference between Bohmian mechanics as reviewed by e.g. Goldstein [38], and de Broglie's original pilot wave theory [86]. This difference is immaterial for my discussion.

Hooft's theory the hidden variable determines the outcomes as well as the settings.[23]
More specifically:

- In Bohmian mechanics the hidden variable is position q, and $d\mu_\psi(q) = |\psi(q)|^2 dq$
  is the Born probability for outcome q with respect to the expansion $|\psi\rangle = \int dq\, \psi(q)|q\rangle$.
- In 't Hooft's theory the hidden variable is a basis vector $|m\rangle$ in some separable Hilbert space H ($m \in \mathbb{N}$), and once again the measure $\mu_\psi(m) = |c_m|^2$ is given by the Born probability for outcome m with respect to the expansion $|\psi\rangle = \sum_m c_m|m\rangle$.

Thus the hidden variables (i.e. $q \in Q$ and $m \in \mathbb{N}$, respectively) have familiar quantum-mechanical interpretations and also their compatibility measures are precisely the Born measures for the quantum state ψ. In this light, we may ask to what extent these hidden variable theories are truly deterministic, as their adherents claim them to be. Since the argument does not rely on entanglement and hence on a bipartite experiment, we may as well work with a quantum coin toss. The settings of the experiments are then fixed, so that we may treat Bohmian mechanics and 't Hooft's theory on the same footing. Idealizing to an infinite run, one has an outcome sequence $x : \mathbb{N} \to \underline{2}$. Standard (Copenhagen) quantum mechanics refuses to say anything about its origin, but nonetheless it does make very specific predictions about x. The basis of these predictions is the following theorem, whose notation and proof are explained in Appendix 1. One may think of a fair quantum coin, in which $\sigma(a) = \underline{2} = \{0, 1\}$ and $\mu_a(0) = \mu_a(1) = 1/2$, and which *probabilistically* is indistinguishable from a fair classical coin (which in my view cannot exist, cf. Sect. 3.5).

Theorem 3.4.1 *The following procedures for repeated identical independent measurements are equivalent (in giving the same possible outcome sequences with the same probabilities):*

1. *Quantum mechanics is applied to the whole run, described as a single quantum-mechanical experiment with a single classically recorded outcome sequence;*
2. *Quantum mechanics is applied to single experiments (with classically recorded outcomes), upon which classical probability theory takes over to combine these.*

Either way, the (purely theoretical) Born probability μ_a for single outcomes induces the infinite Bernoulli process probability μ_a^∞ on the space $\sigma(a)^{\mathbb{N}}$ of infinite outcome sequences.

---

[23] In Bohmian mechanics, the hidden state $q \in Q$ just pertains to the particles undergoing measurement, whilst the settings a are supposed to be "freely chosen" for each measurement (and in particular are independent of q). The outcome is then fixed by a and q. In 't Hooft's theory, the hidden state $x \in X$ of "the world" determines the settings as well as the outcomes. Beyond the issue raised in the main text, Bohmians (but not 't Hooft!) therefore have an additional problem, namely the origin of the settings (which are simply left out of the theory). This weakens their case for determinism even further.

Theorem B.3 in Appendix 2 then implies:

Corollary 3.4.2 *With respect to the "fair" probability measure P^∞ on $\underline{2}^\mathbb{N}$ almost every outcome sequence x of an infinitely often repeated fair quantum coin flip is 1-random.*

In hidden variable theories, on the other hand, x factors through Λ, that is,[24] there are functions $h : \mathbb{N} \to \Lambda$ and $g : \Lambda \to \underline{2}$ such that $x = g \circ h$. Hidden variable theories do provide g, i.e. describe the outcome of any experiment given the value of the hidden variable $\lambda \in \Lambda$. However, what about h, that is, the specification of the value of the hidden variable λ in each run of the experiment? There are just the following two scenarios:

1. *The function h is provided by the hidden variable theory.* In that case, since the theory is supposed to be deterministic, h explicitly gives the values $\lambda_n = h(n)$ for each $n \in \mathbb{N}$ (i.e. experiment no. n in the run). Since g is also given, this means that x is given by the theory. By Theorem B.4 (i.e. Chaitin's second incompleteness theorem), the outcome sequence cannot be 1-random, against Corollary 3.4.2.

2. *The function h is not provided by the hidden variable theory.* In that case, the theory fails to determine the outcome of any specific experiment and just provides averages of outcomes. My conclusion would be that, except for some kind of a "story", nothing has been gained over quantum mechanics, but hidden variable theorists argue that their theories cannot be expected to provide initial conditions (for experiments), and claim that the randomness in measurement outcomes originates in the randomness of the initial conditions of the experiment.[25] But then the question arises what else provides these conditions, and hence our function h. The point here is that in order to recover the predictions of quantum mechanics as meant in Corollary 3.4.2, the function h must sample the Born measure (in its guise of the compatibility measure μ_ψ on Λ), in the sense of "randomly" picking elements from Λ, distributed according to μ_ψ, cf. (3.5.2). This, in turn, should guarantee that the sequences $x = g \circ h$ mimic fair coin flips. Since g is supposed to be given, this implies that the randomness properties of x must entirely originate in h. This origin cannot be deterministic, since in that case we are back to the contradictory scenario 1 above. Hence h must come from some unknown external random process in nature that our hidden variable theories invoke as a kind of an oracle. In my view the need for such a random oracle undermines their purpose and makes them self-defeating. Every way you look at this you lose!

---

[24] The function g *incorporates* all details of the experiment that may affect the outcome (like the setting, context, and quantum state) except the hidden variable λ (which it *specifies*). It has nothing to do with noncontextual value assignments on the set of quantum-mechanical observables (which do not exist).

[25] The Bohmians are divided on the origin of their compatibility measure, referred to in this context as the *quantum equilibrium distribution*, cf. Dürr, Goldstein, and Zanghi [29] against Valentini [86]. The origin of μ_ψ is not my concern, which is the need to randomly sample it and the justification for doing so.

3.5   Conclusion and Discussion

We may summarize the discussion in the previous section as follows[26]:

Theorem 3.5.1 *For any hidden variable theory* T *the following properties are incompatible:*

1. Determinism: T *states the outcome of the measurement of any observable* a given the value $\lambda \in \Lambda$ of the *hidden variable via a function* $g : \Lambda \to \sigma(a)$ and *provides these values for each experiment; for an infinite run this is done via some function* $h : \mathbb{N} \to \Lambda$, *so that* T *provides the outcome sequence* $x : \mathbb{N} \to \sigma(a)$ *through* $x = g \circ h$.
2. Born rule: *Outcome sequences are almost surely 1-random. (cf. Corollary 3.4.2).*

The proof is short. According to the first clause T states the entire outcome sequence x. By Chaitin's incompleteness theorem B.4 this is incompatible with the second clause.   □

In order to understand Theorem 3.5.1 and its proof it may be helpful to note that in classical coin tossing the role of the hidden state is also played by the initial conditions (cf. [25], Chap. 1, Appendix 2). The 50-50 chances (allegedly) making the coin fair are obtained by averaging over the initial conditions, i.e., by sampling. By the same arguments, this sampling cannot be deterministic, i.e. given by a function like h, for otherwise the outcome sequences appropriate to a fair coin would not obtain: it must be done in a genuinely random way and hence by appeal to an external random process. This is impossible classically, so that-unless they have a quantum-mechanical seed-*fair classical coins do not exist*, as confirmed by Diaconis and Skyrms ([25], Chap. 1).

I conclude that deterministic hidden variable theories compatible with quantum mechanics do not exist. The reason that Bell's [2] theorem and the free will theorem leave two loopholes for determinism (i.e. local contextuality and no free choice) is that their compatibility condition with quantum mechanics is stated too weakly: the theory is only required to reproduce certain single-case (Born) probabilities, as opposed to the properties of typical outcome sequences (from which the said probabilities are extracted as long-run frequencies). This reason this approach is still partly successful lies in the clever use of entangled states. If one rejects the second requirement on determinism in Theorem 3.5.1, Bell's theorem and the free will theorem still provide useful constraints on deterministic hidden variable theories, but as shown in the previous section such a rejection necessitates an appeal to an unknown random process and hence seems self-defeating.

Let us now consider the role of the idealization to infinite outcome sequences and see what happens if the experimental runs are finite.[27] Once again, via Theorem 3.4.1

---

[26] In stating the second condition I have taken $\sigma(a) = \{0, 1\}$ with 50-50 Born probabilities, but this can be generalized to other spectra and probability measures. See Downey and Hirschfeldt [27], §6.12.

[27] In other words, we examine whether Earman's principle is satisfied, cf. footnote 37.

the Born rule predicts that outcome strings will be Kolmogorov random with high probability. Any deterministic theory (in the sense of Theorem 3.5.1) provides an explicit description (say in ZFC) of the outcomes, whose randomness would be provable from this description. But this is precluded by Chaitin's first incompleteness theorem (i.e. Theorem B.1), now in the role played by his second incompleteness theorem in the infinite case.[28] □

Nonetheless, although their incompatibility with quantum mechanics has now been established, it will be hard to disprove deterministic hidden variable theories from experimental data. Let us look at the proof of Bell's theorem for inspiration as to what such a (dis)proof should look like. In the context of the EPR–Bohm experiment local deterministic hidden variable theories predict correlations that *satisfy* the Bell inequalities,[29] whereas on suitable settings quantum mechanics predicts (and experiment shows) that typical outcome sequences *violate* these inequalities. Now a disproof of some deterministic hidden variable theory T cannot perhaps be expected to show that *all* quantum-mechanical outcome sequences violate the predictions of the hidden variable theory (indeed they do not, albeit with low probability), but it should identify at least a sufficiently large number of typical (i.e. random) sequences. However, even in the finite case this identification is impossible by Theorem B.1, so that the false predictions of T cannot really be confronted with the correct pre-

---

[28] To make this argument completely rigorous one would need to define what a "description" provided by a deterministic theory means logically. There is a logical characterization of deterministic theories [59], and there are some arguments to the effect that the evolution laws in deterministic theories should be computable, cf. Earman [31], Chap. 11, and Pour-El and Richards [65], *passim*, but this literature makes no direct reference to output strings or sequences of the kind we analyze and in any case the identification of "deterministic" with "computable" is obscure even in situations where the latter concept is well defined. For example, if we stipulate that $h : \mathbb{N} \to \Lambda$ is computable (and likewise $g : \Lambda \to \underline{2}$) then the above appeal to Chaitin's first incompleteness theorem is not even necessary, but this seems too easy. A somewhat circular solution, proposed by Scriven [70], is to simply say that T is deterministic iff the output strings or sequences it describes are not random, but this begs for a more explicit characterization. One might naively expect such a characterization to come from the *arithmetical hierarchy* (found in any book on computability): if, as before, we identify $\underline{2}^{\mathbb{N}}$ with the power set $P(\mathbb{N})$ of $\mathbb{N}$, then $S \subset \mathbb{N}$ is called *arithmetical* if there is a formula $\psi(x)$ in PA (Peano Arithmetic) such that $n \in S$ iff $\mathbb{N} \vDash \psi(n)$, that is, $\psi(n)$ is true in the usual sense. We may then classify the arithmetical subsets through the logical form of ψ, assumed in prenex normal form (i.e., all quantifiers have been moved to the left): S is in $\Sigma_0^0 = \Pi_0^0$ iff ψ has no quantifiers or only bounded quantifiers (in which case S is computable), and then recursively $S \in \Sigma_{n+1}^0$ iff $\psi(x) = \exists_y \varphi(x, y)$ with $\varphi \in \Pi_n^0$, and $\varphi \in \Pi_{n+1}^0$ iff $\psi(x) = \forall_y \varphi(x, y)$ with $\varphi \in \Sigma_n^0$. Here any singly quantified expression $\exists_y \varphi(x, y)$ may be replaced by $\exists y_1 \cdots \exists y_k \varphi(x, y_1, \ldots, y_k)$ and likewise for $\forall_y$. By convention $\Sigma_n^0 \subset \Sigma_{n+1}^0$ and $\Pi_n^0 \subset \Pi_{n+1}^0$, and $\Delta_n^0 := \Sigma_n^0 \cap \Pi_n^0$. Since in classical logic $\forall_y \varphi(x, y)$ is equivalent to $\neg \exists_y \neg \varphi(x, y)$, it follows that Π_n^0 sets are the complements of Σ_n^0 sets. One would then like to locate deterministic theories somewhere in this hierarchy, preferably above the computable Δ_0^0. The idea of a hidden variable (namely y) suggests Σ_1^0 and closure under complementation (it would be crazy if some deterministic theory prefers ones over zeros) then leads to Δ_1^0, but this equals Δ_0^0. The next level Δ_2^0 is impossible since this already contains 1-random sets like Chaitin's Ω. Hence more research is needed.

[29] For Bell's proof it is irrelevant whether or not some hidden variable is able to sample the compatibility measure, since the Bell inequalities follow from pointwise bounds, cf. Landsman [52], Eq. (6.119).

dictions of quantum mechanics. Thus the unprovability of their falsehood condemns deterministic hidden variable theories, and perhaps even determinism as a whole, to a zombie-like existence in a twilight zone comparable with the Dutch situation around selling soft drugs: although this is forbidden by law, it is (officially) not prosecuted.

The situation would change drastically if deterministic hidden variable theories gave up their compatibility with the Born rule (on which my entire reasoning is based), as for example Valentini [86] has argued in case of the de Broglie–Bohm pilot wave theory. For it is this compatibility requirement that kills such theories, which could leave zombie-dom if only they were brave enough to challenge the Born rule. This might open the door to superluminal signaling and worse, but on the other hand the possibility of violating the Born rule would also provide a new context for deriving it, e.g. as a dynamical equilibrium condition (as may be the case for the Broglie–Bohm theory, if Valentini is right).

I would personally expect that the Born rule is emergent from some lower-level theory, which equally well suggests that it is valid in some limit only, rather than absolutely.

The author is grateful to Jacob Barandes, Jeremy Butterfield, Cristian Calude, Erik Curiel, John Earman, Bas Terwijn, and Noson Yanofsky, as well as to members of seminar audiences and especially readers of the first version of this essay on the FQXi website for very helpful comments and corrections. He is even more grateful to the late Michael Redhead, for his exemplary approach to the foundations of physics.

Appendix A: The Born Rule

The Born measure is a probability measure μ_a on the spectrum $\sigma(a)$ of a (bounded) self-adjoint operator a on some Hilbert space H, defined as follows by any state ω on $B(H)$[30]:

Theorem A.1 *Let H be a Hilbert space, let $a^* = a \in B(H)$, and let ω be a state on $B(H)$. There exists a unique probability measure μ_a on the spectrum $\sigma(a)$ of a such that*

$$\omega(f(a)) = \int_{\sigma(a)} d\mu_a(\lambda)\, f(\lambda), \quad \text{for all } f \in C(\sigma(a)). \qquad (3.5.1)$$

The Born measure is a mathematical construction; what is its relationship to experiment? This relationship must be the source of the (alleged) randomness of quantum mechanics, for the Schrödinger equation is deterministic. We start by postulating, as usual, that $\mu_a(\Delta)$ is the (single case) "probability" that measurement of the observable a in the state ω (which jointly give rise to the pertinent Born measure μ_a) gives a result

---

[30] Here a state ω is a positive normalized linear functional on $B(H)$, as in the C*-algebraic approach to quantum mechanics (Haag [41], Landsman [52]). One may think of expectation values $\omega(a) = \mathrm{Tr}(\rho a)$, where ρ is a density operator on H, with the special case $\omega(a) = \langle \psi, a\psi \rangle$, where $\psi \in H$ is a unit vector. For a proof of Theorem A.1 see Landsman [52], Sect. 4.1, Corollary 4.4.

$\lambda \in \Delta \subset \sigma(a)$. Here I identify single-case "probabilities" with numbers (consistent with the probability calculus) provided by *theory*, upon which long-run frequencies provide *empirical evidence* for the theory in question, but do not *define* probabilities. The Born measure is a case in point: these probabilities are *theoretically given*, but have to be *empirically verified* by long runs of independent experiments. In other words, by the results reviewed below such experiments provide numbers whose role it is to test the Born rule as a hypothesis. This is justified by the following **sampling theorem** (strong law of large numbers): for any (measurable) subset $\Delta \subset \sigma(a)$ and any sequence $(x_n) \in \sigma(a)^{\mathbb{N}}$ we have μ_a^∞-almost surely:

$$\lim_{N \to \infty} \frac{1}{N}(1_\Delta(x_1) + \cdots + 1_\Delta(x_N)) = \mu_a(\Delta). \tag{3.5.2}$$

Proof of Theorem 3.4.1. Let $a = a^* \in B(H)$, where H is a Hilbert space and $B(H)$ is the algebra of all bounded operators on H, and let $\sigma(a)$ be the spectrum of a. For simplicity (and since this is enough for our applications, where $H = \mathbb{C}^2$) I assume $\dim(H) < \infty$, so that $\sigma(a)$ simply consists of the eigenvalues λ_i of a (which may be degenerate). Let us first consider a *finite* number N of identical measurements of a (a "run"). The first option in the theorem corresponds to a simultaneous measurement of the commuting operators

$$a_1 = a \otimes 1_H \otimes \cdots \otimes 1_H; \tag{3.5.3}$$

$$\cdots$$

$$a_N = 1_H \otimes \cdots \otimes 1_H \otimes a, \tag{3.5.4}$$

all defined on the N-fold tensor product $H^N \equiv H^{\otimes N}$ of H with itself.[31] To put this in a broader perspective, consider *any* set $(a_1, \ldots, a_N) \equiv \underline{a}$ of commuting operators on *any* Hilbert space K (of which (3.5.3)–(3.5.4) is obviously a special case with $K = H^N$). These operators have a *joint spectrum* $\sigma(\underline{a})$, whose elements are the *joint eigenvalues* $\underline{\lambda} = (\lambda_1, \ldots, \lambda_N)$, defined by the property that there exists a nonzero joint eigenvector $\psi \in K$ such that $a_i \psi = \lambda_i \psi$ for all $i = 1, \ldots, N$; clearly,

$$\sigma(\underline{a}) = \{\underline{\lambda} \in \sigma(a_1) \times \cdots \times \sigma(a_n) \mid e_{\underline{\lambda}} \equiv e_{\lambda_1}^{(1)} \cdots e_{\lambda_n}^{(n)} \neq 0\} \subseteq \sigma(a_1) \times \cdots \times \sigma(a_N), \tag{3.5.5}$$

where $e_{\lambda_i}^{(i)}$ is the spectral projection of a_i on the eigenspace for the eigenvalue $\lambda_i \in \sigma(a_i)$. Von Neumann's Born rule for the probability of finding $\underline{\lambda} \in \sigma(\underline{a})$ then simply reads

$$p_{\underline{a}}(\underline{\lambda}) = \omega(e_{\underline{\lambda}}), \tag{3.5.6}$$

where ω is the state on $B(K)$ with respect to which the Born probability is defined.[32] If $\dim(K) < \infty$, as I assume, we always have $\omega(a) = \mathrm{Tr}\,(\rho a)$ for some density operator ρ, and for a general Hilbert space K this is the case iff the state ω is normal

---

[31] This can even be replaced by a single measurement, see Landsman [52], Corollary A.20.

[32] The uses of states themselves may be justified by Gleason's theorem ([52], §§2,7, 4.4).

on $B(K)$. For (normal) pure states we have $\rho = |\psi\rangle\langle\psi|$ for some unit vector $\psi \in K$, in which case

$$p_{\underline{a}}(\underline{\lambda}) = \langle\psi, e_{\underline{\lambda}}\psi\rangle. \tag{3.5.7}$$

The Born rule (3.5.6) is similar to the single-operator case ([52], §4.1)[33]: the continuous functional calculus gives a Gelfand isomorphism of commutative C*-algebras

$$C^*(\underline{a}, 1_K) \cong C(\sigma(\underline{a})), \tag{3.5.8}$$

under which the restriction of the state ω, originally defined on $B(K)$, to its commutative C*-subalgebra $C^*(\underline{a})$ defines a probability measure $\mu_{\underline{a}}$ on the joint spectrum $\sigma(\underline{a})$ via the Riesz isomorphism. This is the Born measure, whose probabilities are given by (3.5.6). For the case (3.5.3)–(3.5.4) we have equality in (3.5.5); since in that case $\sigma(a_i) = \sigma(a)$, we obtain

$$\sigma(\underline{a}) = \sigma(a)^N, \tag{3.5.9}$$

and therefore, for all $\lambda_i \in \sigma(a)$ and states ω on $B(H^N)$, the Born rule (3.5.6) becomes

$$p_{\underline{a}}(\lambda_1, \ldots, \lambda_N) = \omega(e_{\lambda_1} \otimes \cdots \otimes e_{\lambda_N}). \tag{3.5.10}$$

Now take a state ω_1 on $B(H)$. Reflecting the idea that ω is the state on $B(H^N)$ in which N independent measurements of $a \in B(H)$ in the state ω_1 are carried out, choose

$$\omega = \omega_1^N, \tag{3.5.11}$$

the state on $B(H^N)$ defined by linear extension of its action on elementary tensors:

$$\omega_1^N(b_1 \otimes \cdots \otimes b_n) = \omega_1(b_1) \cdots \omega_N(b_N). \tag{3.5.12}$$

It follows that

$$\omega^N(e_{\lambda_1} \otimes \cdots \otimes e_{\lambda_N}) = \omega_1(e_{\lambda_1}) \cdots \omega_1(e_{\lambda_N}) = p_a(\lambda_1) \cdots p_a(\lambda_N), \tag{3.5.13}$$

so that the joint probability of the outcome $(\lambda_1, \ldots, \lambda_N) \in \sigma(\underline{a})$ is simply

$$p_{\underline{a}}(\lambda_1, \ldots, \lambda_N) = p_a(\lambda_1) \cdots p_a(\lambda_N). \tag{3.5.14}$$

Since these are precisely the probabilities for option 2 (i.e. the Bernoulli process), i.e.,

$$\mu_{\underline{a}} = \mu_a^N, \tag{3.5.15}$$

---

[33] The Born rule for commuting operators follows from the single operator case ([52], §2.5).

this proves the claim for $N < \infty$. To describe the limit $N \to \infty$, let B be any C*-algebra with unit 1_B; below I take $B = B(H)$, $B = C^*(a, 1_H)$, or $B = C(\sigma(a))$. We now take

$$A_N = B^{\otimes N}, \tag{3.5.16}$$

the N-fold tensor product of B with itself.[34] The special cases above may be rewritten as

$$B(H)^{\otimes N} \cong B(H^N); \tag{3.5.17}$$

$$C^*(a, 1_H)^{\otimes N} \cong C^*(a_1, \ldots, a_N, 1_{H^N}); \tag{3.5.18}$$

$$C(\sigma(a))^{\otimes N} \cong C(\sigma(a) \times \cdots \times \sigma(a)), \tag{3.5.19}$$

with N copies of H and $\sigma(a)$, respectively, and in (3.5.18) the a_i are given by (3.5.3)–(3.5.4). We may then wonder if these algebras have a limit as $N \to \infty$. They do, but it is not unique and depends on the choice of observables, that is, of the infinite sequences $\mathbf{a} = (a_1, a_2, \ldots)$, with $a_N \in A_N$, that are supposed to have a limit. One possibility is to take sequences $\mathbf{a}$ for which there exists $M \in \mathbb{N}$ and $a_M \in A_M$ such that for each $N \geq M$,

$$a_N = a_M \otimes 1_B \cdots \otimes 1_B, \tag{3.5.20}$$

with $N - M$ copies of 1_B. On that choice, one obtains the infinite tensor product $B^{\otimes \infty}$, see Landsman [52], §C.14. The limit of (3.5.17) in this sense is $B(H^{\otimes \infty})$, where $H^{\otimes \infty}$ is von Neumann's 'complete' infinite tensor product of Hilbert spaces,[35] in which $C^*(a, 1_H)^{\otimes \infty}$ is the C*-algebra generated by $(a_1, a_2, \ldots)$ and the unit on $H^{\otimes \infty}$. The limit of (3.5.19) is

$$C(\sigma(a))^{\otimes \infty} \cong C(\sigma(a)^{\mathbb{N}}), \tag{3.5.21}$$

where $\sigma(a)^{\mathbb{N}}$, which we previously saw as a measure space (as a special case of $X^{\mathbb{N}}$ for general compact Hausdorff spaces X), is now seen as a topological space with the product topology, in which it is compact.[36] As in the finite case, we have an isomorphism

$$C^*(a, 1_H)^{\otimes \infty} \cong C(\sigma(a))^{\otimes \infty}, \tag{3.5.22}$$

and hence, on the given identifications, we obtain an isomorphism of C*-algebras

$$C^*(a_1, a_2, \ldots, 1_{H^{\otimes \infty}}) \cong C(\sigma(a)^{\mathbb{N}}). \tag{3.5.23}$$

---

[34] If B is infinite-dimensional, for technical reasons the so-called *projective* tensor product should be used.

[35] See Landsman [52], §8.4 for this approach. The details are unnecessary here.

[36] Cf. Tychonoff's theorem. The associated Borel structure is the one defined by the cylinder sets.

It follows from the definition of the infinite tensor products used here that each state ω_1 on B defines a state ω_1^∞ on $B^{\otimes\infty}$. Take $B = B(H)$ and restrict ω_1^∞, which a priori is a state on $B(H^{\otimes\infty})$, to its commutative C*-subalgebra $C^*(a_1, a_2, \ldots, 1_{H^{\otimes\infty}})$. The isomorphism (3.5.23) then gives a probability measure $\mu_{\underline{a}}$ on the compact space $\sigma(a)^{\mathbb{N}}$, where the label $\underline{a}$ now refers to the infinite set of commuting operators $(a_1, a_2, \ldots)$ on $H^{\otimes\infty}$. To compute this measure, I use (3.5.1) and the fact that by construction functions of the type

$$f(\lambda_1, \lambda_2, \ldots) = f^{(N)}(\lambda_1, \ldots, \lambda_N), \tag{3.5.24}$$

where $N < \infty$ and $f^{(N)} \in C(\sigma(a)^N)$, are dense in $C(\sigma(a)^{\mathbb{N}})$ (with respect to the appropriate supremum-norm), and that in turn finite linear combinations of factorized functions $f^{(N)}(\lambda_1, \ldots, \lambda_N) = f_1(\lambda_1) \cdots f_N(\lambda_N)$ are dense in $C(\sigma(a)^N)$. It follows from this that

$$\mu_{\underline{a}} = \mu_a^\infty. \tag{3.5.25}$$

Since this generalizes (3.5.15) to $N = \infty$, the proof of Theorem 3.4.1 is finished. $\square$

Appendix B: 1-Randomness

In what follows, the notion of 1-randomness, originally defined by Martin-Löf in the setting of constructive measure theory, will be explained through an equivalent definition in terms of Kolmogorov complexity.[37] We assume basic familiarity with the notion of a computable function $f : \mathbb{N} \to \mathbb{N}$, which may be defined through recursion theory or Turing machines.

A *string* is a *finite* succession of bits (i.e. zeros and ones). The length of a string σ is denoted by $|\sigma|$. The set of all strings of length N is denoted by $\underline{2}^N$, where $\underline{2} = \{0, 1\}$, and

$$\underline{2}^* = \bigcup_{N \in \mathbb{N}} \underline{2}^N \tag{3.5.26}$$

denotes the set of all strings. The *Kolmogorov complexity* $K(\sigma)$ of $\sigma \in \underline{2}^*$ is defined, roughly speaking, as the length of the shortest computer program that prints σ and then halts. We then say, again roughly, that σ is *Kolmogorov random* if this shortest program contains all of σ in its code, i.e. if the shortest computable description of σ is σ itself.

---

[37] For details see Volchan [87], Terwijn [85], Diaconis and Skyrms ([25], Chap. 8), and Eagle [30] for starters, technical surveys by Zvonkin and Levin [93], Muchnik et al. [61], Downey et al. [28], Grünwald and Vitányi [40], and Dasgupta [24], and books by Calude [15], Li and Vitányi [56], Nies [63], and Downey and Hirschfeldt [27]. For history see van Lambalgen [50, 51] and Li and Vitányi [56]. For physical applications see e.g. Earman [31], Svozil [77, 78], Calude [16], Wolf [90], Bendersky et al. [4, 5], Senno [73], Baumeler et al. [1], and Tadaki [81, 82].

To make this precise,[38] fix some universal prefix-free Turing machine U, seen as performing a computation on input τ (in its prefix-free domain) with output $U(\tau)$, and define

$$K(\sigma) = \min_{\tau \in \underline{2}^*}\{|\tau| : U(\tau) = \sigma\}. \qquad (3.5.27)$$

The function $K : \underline{2}^* \to \mathbb{N}$ is uncomputable, but that doesn't mean it is ill-defined. The choice of U affects $K(\sigma)$ up to a σ-independent constant, and to take this dependency into account we state certain results in terms of the "big-O" notation familiar from Analysis.[39] For example, if σ is easily computable, like the first $|\sigma|$ binary digits of π, then

$$K(\sigma) = O(\log|\sigma|), \qquad (3.5.28)$$

with the logarithm in base 2 (as only the length of σ counts). However, a random σ has

$$K(\sigma) = |\sigma| + O(\log|\sigma|). \qquad (3.5.29)$$

We say that σ is *c-Kolmogorov random*, for some σ-independent constant $c \in \mathbb{N}$, if

$$K(\sigma) \geq |\sigma| - c. \qquad (3.5.30)$$

Simple counting arguments show that as $|\sigma| = N$ gets large, the overwhelming majority of strings in $\underline{2}^N$ (and hence in $\underline{2}^*$) is c-random.[40] The following theorem, which might be called *Chaitin's first incompleteness theorem*, therefore shows that randomness is elusive[41]:

---

[38] A Turing machine T is *prefix-free* if its domain $D(T)$ consists of a prefix-free subset of $\underline{2}^*$, i.e., if $\sigma \in D(T)$ then $\sigma\tau \notin D(T)$ for any $\sigma, \tau \in \underline{2}^*$, where $\sigma\tau$ is the concatenation of σ and τ: if T halts on input σ then it does not halt on either any initial part or any extension of σ. The prefix-free version is only needed to correctly define randomness of sequences in terms of randomness of their initial parts, which is necessary to satisfy *Earman's Principle*: 'While idealizations are useful and, perhaps, even essential to progress in physics, a sound principle of interpretation would seem to be that no effect can be counted as a genuine physical effect if it disappears when the idealizations are removed.' See Earman [32], p. 191. For finite strings σ one may work with the *plain Kolmogorov complexity* $C(\sigma)$, defined as the length (in bits) of the shortest computer program (run on some fixed universal Turing machine U) that computes σ.

[39] Recall that $f(n) = O(g(n))$ iff there are constants C and N such that $|f(n)| \leq C|g(n)|$ for all $n \geq N$.

[40] It is easy to show that least $2^N - 2^{N-c+1} + 1$ strings σ of length $|\sigma| = N$ are c-Kolmogorov random.

[41] Here "sound" means that all theorems proved by T are true; this is a stronger assumption than consistency (in fact only the arithmetic fragment of T needs to be sound). One may think of Peano Arithmetic (PA) or of Zermelo–Fraenkel set theory with the axiom of choice (ZFC). As in Gödel's theorems, one also assumes that T is formalized as an axiomatic-deductive system in which proofs could in principle be carried out mechanically by a computer. The status of the true but unprovable sentences $K(\sigma) > C$ in Chaitin's theorem is similar to that of the sentence G in Gödel's original proof of his first incompleteness theorem, which roughly speaking is an arithmetization of the statement "I cannot be proved in T": assuming soundness and hence consistency of T, one can prove G and $K(\sigma) > C$ in the usual interpretation of the arithmetic fragment of T in the natural

Theorem B.1 *For any sound mathematical theory T containing enough arithmetic there is a constant $C \in \mathbb{N}$ such that T cannot prove any sentence of the form $K(\sigma) > C$ (although infinitely many such sentences are true), and as such T can only prove (Kolmogorov) randomness of finitely many strings (although infinitely many strings are in fact random).*

The proof is quite complicated in its details but it is based on the existence of a computably enumerable (c.e.) list $\mathsf{T} = (\tau_1, \tau_2, \ldots)$ of the theorems of T, and on the fact that after Gödelian encoding by numbers, theorems of any given grammatical form can be computably searched for in this list and will eventually be found. In particular, there exists a program P (running on the universal prefix-free Turing machine U used to define $K(\cdot)$) such that $P(n)$ halts iff there exists a string σ for which $K(\sigma) > n$ is a theorem of T. If there is such a theorem the output is $P(n) = \sigma$, where σ appears in the first such theorem of the kind (according to the list T). By definition of $K(\cdot)$, this means that

$$K(\sigma) \leq |P| + |n|. \tag{3.5.31}$$

Now suppose that no C as in the above statement of the theorem exists. Then there is $n \in \mathbb{N}$ large enough that $n > |P| + |n|$ and there is a string $\sigma \in 2^*$ such that T proves $K(\sigma) > n$. Since T is sound this is actually true,[42] which gives a contradiction between

$$K(\sigma) > n > |P| + |n|; \qquad\qquad K(\sigma) \leq |P| + |n|. \tag{3.5.32}$$

Note that this proof shows that a *proof* in T of $K(\sigma) > n$ (if true) would also *identify* σ.

As an idealization of a long (binary) string, a (binary) *sequence* $x = x_1 x_2 \cdots$ is an infinite succession of bits, i.e. $x \in 2^{\mathbb{N}}$, with finite truncations $x_{|N} = x_1 \cdots x_N \in 2^N$ for each $N \in \mathbb{N}$. We then call x *Levin–Chaitin random* if each truncation of x is c-Kolmogorov random for some c, that is, if there exists $c \in \mathbb{N}$ such that

$$K(x_{|N}) \geq N - c$$

for each $N \in \mathbb{N}$. Equivalently,[43] a sequence x is Levin–Chaitin random if eventually $K(x_{|N}) >> N$, in that

$$\lim_{N \to \infty} (K(x_{|N}) - N) = \infty. \tag{3.5.33}$$

---

numbers $\mathbb{N}$. See Chaitin [20] for his own presentation and analysis of his incompleteness theorem. Raatikainen [66] also gives a detailed presentation of the theorem, including a critique of Chaitin's ideology. Incidentally, he shows that there even exists a U with respect to which $K(\cdot)$ is defined such that $C = 0$ in ZFC. See also Franzén [35] and Gács [36].

[42] The following contradiction can be made more dramatic by taking n such that $n >> |P| + |n|$.

[43] See Calude [15], Theorem 6.38 (attributed to Chaitin) for this equivalence.

Apart from having the same intuitive pull as Kolmogorov randomness (of strings), this definition gains from the fact that it is equivalent to two other appealing notions of randomness, namely *patternlessness* and *unpredictability*, both also defined computationally. In view of these equivalences we simply call a Levin–Chaitin random sequence *1-random*.[44]

A sequence $x \in \underline{k}^{\mathbb{N}}$ is *Borel normal* in base k if each string σ has frequency $k^{-|\sigma|}$ in x. Any hope of defining randomness as Borel normality in base 10 is blocked by *Champernowne's number* $0123456789101112131\ldots$, which is Borel normal but clearly not random in any reasonable sense (this is also true in base 2). The decimal expansion of π is also conjectured to be Borel normal in base 10 (with huge numerical support), although π clearly is not random either. However, Borel normality seems a desirable property of truly random numbers on any good definition, and so we are fortunate to have:

Proposition B.2 *A 1-random sequence is Borel normal (in base 2, but in fact in any base) and hence ("monkey typewriter theorem") contains any finite string infinitely often.*[45]

Another desirable property comes from the following theorem due to Martin-Löf, in which P is the 50-50 probability on $\{0, 1\}$ and P^{∞} is the induced probability measure on $\underline{2}^{\mathbb{N}}$:

Theorem B.3 *With respect to P^{∞} almost every outcome sequence $x \in \underline{2}^{\mathbb{N}}$ is 1-random.*

This implies that the 1-random sequences form an uncountable subset of $\underline{2}^{\mathbb{N}}$,[46] although topologically this subset is meagre (i.e. Baire first category).[47] Chaitin's incompleteness theorem for (finite) strings has the following counterpart for (infinite) sequences:

---

[44] Any pattern in a sequence x would make it compressible, but one has to define the notion of a pattern very carefully in a computational setting. This was accomplished by Martin-Löf in 1966, who defined a pattern as a specific kind of probability-zero subset T of $\underline{2}^{\mathbb{N}}$ (called a "test") that can be computably approximated by subsets $T_n \subset \underline{2}^{\mathbb{N}}$ of increasingly small probability 2^{-n}; if $x \in T$, then x displays some pattern and it is patternless iff $x \notin T$ for all such tests. Martin-Löf's definition yields what usually called 1-randomness, in view of his use of so-called Σ_1^0 sets. See the textbooks Li and Vitányi [56], Calude [15], Nies [63], and Downey and Hirschfeldt [27] for the equivalences between Levin–Chaitin randomness (incompressibility), Martin-Löf randomness (patternlessness), and a third notion (unpredictability) that evolved from the work of von Mieses and Ville, finalized by Schnorr. The name *Levin–Chaitin randomness*, taken from Downey et al. [28], is justified by its independent origin in Levin [55] and Chaitin [19].

[45] For details and proofs see Calude [15], Corollary 6.32 in §6.3 and almost all of §6.4.

[46] To see this, use the measure-theoretic isomorphism between $(\underline{2}^{\mathbb{N}}, \Sigma_K, P^{\infty})$ and $([0, 1], \Sigma_L, dx)$, where Σ_K is the "Kolmogorov" σ-algebra generated by the cylinder sets $[\sigma] = \{x \in \underline{2}^{\mathbb{N}} \mid x_{\|\sigma\|} = \sigma\}$, where $\sigma \in \underline{2}^*$, and Σ_K is the "Lebesgue" σ-algebra generated by the open subsets of $[0, 1]$. See also Nies [63], §1.8.

[47] See Calude [15], Theorem 6.63. Hence meagre subsets of $[0, 1]$ exist with unit Lebesgue measure!

Theorem B.4 *If $x \in \underline{2}^{\mathbb{N}}$ is 1-random, then ZFC (or any sufficiently comprehensive mathematical theory T meant in Theorem B.1) can compute only finite many digits of x.*[48]

This clearly excludes defining a 1-random number by somehow listing its digits, but some can be described by a formula. One example is Chaitin's Ω, or more precisely Ω_U,[49] which is the halting probability of some fixed universal prefix-free Turing machine U, given by

$$\Omega_U := \sum_{\tau \in \underline{2}^* | U(\tau)\downarrow} 2^{-|\tau|}. \tag{3.5.34}$$

Appendix C: Bell's Theorem and Free Will Theorem

In support of the analysis of hidden variable theories in the main text, this appendix reviews Bell's [2] theorem and the free will theorem, streamlining earlier expositions ([17], [52], Chap. 6) and leaving out proofs and other adornments.[50] In the specific context of 't Hooft's theory (where the measurement settings are determined by the hidden state) and Bohmian mechanics (where they are not, as in the original formulation of Bell's theorem and in most hidden variable theories) an advantage of my approach is that both free (uncorrelated) und correlated settings fall within its scope; the former are distinguished from the latter by an independence assumption.[51]

---

[48] More precisely, only finitely many true statements of the form: 'the n'th bit x_n of x equals its actual value' (i.e. 0 or 1) are provable in T (where a proof in T may be seen as a computation, since one may algorithmically search for this proof in a list). See Calude [15], Theorem 8.7, which is stated for Chaitin's Ω but whose proof holds for any 1-random sequence. Indeed, as pointed out to the author by Bas Terwijn, even more generally, ZFC (etc.) can only compute finitely many digits of any *immune* sequence (we say that a sequence $x \in \underline{2}^{\mathbb{N}}$ is *immune* if the corresponding subset $S \subset \mathbb{N}$ (i.e. $1_S = x$) contains no infinite c.e. subset), and by (for example) Corollary 6.42 in Calude [15] any 1-random sequence is immune.

[49] There exists a U for which not a single digit of Ω_U can be known, see Calude [15], Theorem 8.11.

[50] The original reference for Bell's theorem is Bell [2]; see further footnote 6, and in the context of this appendix also Esfeld [34] and Sen and Valentini [72] are relevant. The free will theorem originates in Heywood and Redhead [44], followed by Stairs [76], Brown and Svetlichny [11], Clifton [18], and, as name-givers, Conway and Kochen [21]. Both theorems can and have been presented and interpreted in many different ways, of which we choose the one that is relevant for the general discussion on randomness in the main body of the paper. This appendix is taken almost *verbatim* from Landsman [53].

[51] This addresses a problem Bell faced even according to some of his most ardent supporters [64, 71], namely the tension between the idea that the hidden variables (in the pertinent causal past) should on the one hand include all ontological information relevant to the experiment, but on the other hand should leave Alice and Bob free to choose any settings they like. Whatever its ultimate fate, 't Hooft's staunch determinism has drawn attention to issues like this, as has the free will theorem.

As a warm-up I start with a version of the Kochen–Specker theorem, whose logical form is very similar to Bell's [2] theorem and the free will theorem, as follows:

Theorem C.1 *Determinism,* QM, *non-contextuality, and free choice are contradictory.*

Of course, this unusual formulation hinges on the precise meaning of these terms.

- ***determinism*** is the conjunction of the following two assumptions.
  1. There is a state space X with associated functions $A : X \to S$ and $L : X \to O$, where S is the set of all possible *measurement settings* Alice can choose from, namely a suitable finite set of orthonormal bases of $\mathbb{R}^3$ (11 well-chosen bases will do to arrive at a contradiction),[52] and O is some set of possible *measurement outcomes*. Thus some $x \in X$ determines *both* Alice's setting $a = A(x)$ *and* her outcome $\alpha = L(x)$.
  2. There exists some set Λ and an additional function $H : X \to \Lambda$ such that

$$L = L(A, H), \qquad (3.5.35)$$

in the sense that for each $x \in X$ one has $L(x) = \hat{L}(A(x), H(x))$ for a certain function $\hat{L} : S \times \Lambda \to O$. This self-explanatory assumption just states that each measurement outcome $L(x) = \hat{L}(a, \lambda)$ is determined by the measurement setting $a = A(x)$ and the "hidden" variable or state $\lambda = H(x)$ of the particle undergoing measurement.
- QM fixes $O = \{(0, 1, 1), (1, 0, 1), (1, 1, 0)\}$, which is a non-probabilistic fact of quantum mechanics with overwhelming (though indirect) experimental support.
- ***non-contextuality*** stipulates that the function $\hat{L}$ just introduced take the form

$$\hat{L}((\vec{e}_1, \vec{e}_2, \vec{e}_3), \lambda) = (\tilde{L}(\vec{e}_1, \lambda), \tilde{L}(\vec{e}_2, \lambda), \tilde{L}(\vec{e}_3, \lambda)), \qquad (3.5.36)$$

for a single function $\tilde{L} : S^2 \times \Lambda \to \{0, 1\}$ that also satisfies $\tilde{L}(-\vec{e}, \lambda) = \tilde{L}(\vec{e}, \lambda)$.[53]
- ***free choice*** finally states that the following function is surjective:

$$A \times H : X \to S \times \Lambda; \qquad x \mapsto (A(x), H(x)). \qquad (3.5.37)$$

In other words, for each $(a, \lambda) \in S \times \Lambda$ there is an $x \in X$ for which $A(x) = a$ and $H(x) = \lambda$. This makes A and H "independent" (or: makes a and λ free variables).

---

[52] If her setting is a basis $(\vec{e}_1, \vec{e}_2, \vec{e}_3)$, Alice measures the quantities $(J_{\vec{e}_1}^2, J_{\vec{e}_2}^2, J_{\vec{e}_3}^2)$, where $J_{\vec{e}_1} = \langle \vec{J}, \vec{e}_i \rangle$ is the component of the angular momentum operator $\vec{J}$ of a massive spin-1 particle in the direction $\vec{e}_i$.

[53] Here $S^2 = \{(x, y, z) \in \mathbb{R}^3 \mid x^2 + y^2 + z^2 = 1\}$ is the 2-sphere, seen as the space of unit vectors in $\mathbb{R}^3$. Equation (3.5.36) means that the outcome of Alice's measurement of $J_{\vec{e}_i}^2$ is independent of the "context" $(J_{\vec{e}_1}^2, J_{\vec{e}_2}^2, J_{\vec{e}_3}^2)$; she might as well measure $J_{\vec{e}_i}^2$ by itself. The last equation is trivial, since $(J_{-\vec{e}_i})^2 = (J_{\vec{e}_i})^2$.

See Landsman [52], §6.2 for a proof of the Kochen–Specker theorem in this language.[54]

Bell's [2] theorem and the free will theorem both take a similar generic form, namely:

Theorem C.2 *Determinism,* QM, *local contextuality, and free choice, are contradictory.*

Once again, I have to explain what these terms exactly mean in the given context.

- *determinism* is a straightforward adaptation of the above meaning to the bipartite "Alice and Bob" setting. Thus we have a state space X with associated functions

$$A : X \to S; \qquad B : X \to S; \qquad L : X \to O \qquad R : X \to O, \qquad (3.5.38)$$

where S, the set of all possible measurement settings Alice and Bob can each choose from, differs a bit between the two theorems: for the free will theorem it is the same as for the Kochen–Specker theorem above, as is the set O of possible measurement outcomes, whereas for Bell's theorem (in which Alice and Bob each measure a 2-level system), S is some finite set of angles (three is enough), and $O = \{0, 1\}$.

  - In the free will case, these functions and the state $x \in X$ determine both the settings $a = A(x)$ and $b = B(x)$ of a measurement and its outcomes $\alpha = L(x)$ and $\beta = R(x)$ for Alice on the *L*eft and for Bob on the *R*ight, respectively.
  - All of this is also true in the Bell case, but since his theorem relies on impossible measurement statistics (as opposed to impossible individual outcomes), one in addition assumes a probability measure μ on X.[55]

Furthermore, there exists some set Λ and some function $H : X \to \Lambda$ such that

$$L = L(A, B, H); \qquad\qquad R = R(A, B, H), \qquad (3.5.39)$$

in the sense that for each $x \in X$ one has functional relationships

$$L(x) = \hat{L}(A(x), B(x), H(x)); \qquad R(x) = \hat{R}(A(x), B(x), H(x)), \qquad (3.5.40)$$

for certain functions $\hat{L} : S \times S \times \Lambda \to O$ and $\hat{R} : S \times S \times \Lambda \to O$.

---

[54] The assumptions imply the existence of a coloring $C_\lambda : \mathcal{P} \to \{0, 1\}$ of $\mathbb{R}^3$, where $\mathcal{P} \subset S^2$ consist of all unit vectors contained in all bases in S, and λ "goes along for a free ride". A coloring of $\mathbb{R}^3$ is a function $C : \mathcal{P} \to \{0, 1\}$ such that for any set $\{e_1, e_2, e_3\}$ in $\mathcal{P}$ with $e_i e_j = \delta_{ij} 1_3$ and $e_1 + e_2 + e_3 = 1_3$ where 1_3 is the 3×3 unit matrix) there is exactly one e_i for which $C(e_i) = 1$. Indeed, one finds $C_\lambda(\bar{e}) = \hat{L}(\bar{e}, \lambda)$. The key to the proof of Kochen–Specker is that on a suitable choice of the set S such a coloring cannot exist.

[55] The existence of μ is of course predicated on X being a measure space with corresponding σ-algebra of measurable subsets, with respect to which all functions in (3.5.38) and below are measurable.

- QM reflects elementary quantum mechanics of correlated 2-level and 3-level quantum systems for the Bell and the free will cases, respectively, as follows[56]:

  - In the *free will theorem*, $O = \{(0, 1, 1), (1, 0, 1), (1, 1, 0)\}$ is the same as for the Kochen–Specker theorem. In addition *perfect correlation* obtains: if $a = (\vec{e}_1, \vec{e}_2, \vec{e}_3)$ is Alice's orthonormal basis and $b = (\vec{f}_1, \vec{f}_2, \vec{f}_3)$ is Bob's, one has

$$\vec{e}_i = \vec{f}_j \Rightarrow \hat{L}_i(a, b, z) = \hat{R}_j(a, b, z), \qquad (3.5.41)$$

  where $\hat{L}_i, \hat{R}_j : S \times S \times \Lambda \to \{0, 1\}$ are the components of $\hat{L}$ and $\hat{R}$, respectively. Finally,[57] if (a', b') differs from (a, b) by changing the sign of any basis vector,

$$\hat{L}(a', b', \lambda) = \hat{L}(a, b, \lambda); \qquad \hat{R}(a', b', \lambda) = \hat{R}(a, b, \lambda). \qquad (3.5.42)$$

  - In *Bell's theorem*, $O = \{0, 1\}$, and the statistics for the experiment is reproduced as conditional joint probabilities given by the measure μ through

$$P(L \neq R | A = a, B = b) = \sin^2(a - b). \qquad (3.5.43)$$

- *local contextuality*, which replaces and weakens non-contextuality, means that

$$L(A, B, H) = L(A, H); \qquad R(A, B, H) = G(B, H). \qquad (3.5.44)$$

  In words: Alice's outcome *given* λ does not depend on Bob's setting, and *vice versa*.
- *free choice* is an independence assumption that looks differently for both theorems:

  - In the *free will theorem* it means that each $(a, b, \lambda) \in S \times S \times \Lambda$ is possible in that there is an $x \in X$ for which $A(x) = a$, $B(x) = b$, and $H(x) = \lambda$.
  - In *Bell's theorem*, (A, B, H) are *probabilistically independent* relative to μ.[58]

This concludes the joint statement of the free will theorem and Bell's [2] theorem in the form we need for the main text. The former is proved by reduction to the Kochen–Specker theorem, whilst the latter follows by reduction to the usual version of Bell's theorem via the free choice assumption; see Landsman [52], Chap. 6 for details.

For our purposes these theorems are equivalent, despite subtle differences in their assumptions. Bell's theorem is much more robust in that it does not rely on perfect correlations (which are hard to realize experimentally), and in addition it requires almost no input from quantum theory. On the other hand, Bell's theorem

---

[56] In Bell's theorem quantum theory can be replaced by experimental support [43].

[57] As in Kochen–Specker, this is because Alice and Bob measure *squares* of (spin-1) angular momenta.

[58] By definition, this also implies that the pairs (A, B), (A, H), and (B, H) are also independent.

uses probability theory in a highly nontrivial way: like the hidden variable theories it is supposed to exclude it relies on the possibility of fair sampling of the probability measure μ. The factorization condition defining probabilistic independence passes this requirement of fair sampling on to both the hidden variable and the settings, which brings us back to the main text.

Different parties may now be identified by the assumption they drop: Copenhagen quantum mechanics rejects determinism, Valentini [86] rejects the Born rule and hence QM, Bohmians rejects local contextuality, and finally 't Hooft rejects free choice. However, as we argue in the main text, even the latter two camps do not really have a deterministic theory underneath quantum mechanics because of their need to randomly sample the probability measure they must use to recover the predictions of quantum mechanics.

References

1. A. Baumeler, C.A. Bédard, G. Brassard, S. Wolf, Kolmogorov amplification from Bell correlation, in *IEEE International Symposium on Information Theory (ISIT)* (2017). https://doi.org/10.1109/ISIT.2017.8006790
2. J.S. Bell, On the Einstein–Podolsky–Rosen paradox. Physics **1**, 195–200 (1964)
3. J.S. Bell, On the problem of hidden variables in quantum mechanics. Rev. Mod. Phys. **38**, 447–452 (1966)
4. A. Bendersky, G. de la Torre, G. Senno, S. Figueira, A. Acín, Algorithmic pseudorandomness in quantum setups. Phys. Rev. Lett. **116**, 230402 (2016)
5. A. Bendersky, G. Senno, G. de la Torre, S. Figueira, A. Acín, Nonsignaling deterministic models for nonlocal correlations have to be incomputable. Phys. Rev. Lett. **118**, 130401 (2017)
6. D. Bohm, A suggested interpretation of the quantum theory in terms of 'hidden' variables, I, II. Phys. Rev. **85**, 166–179, 180–193 (1952)
7. N. Bohr, Can quantum-mechanical description of physical reality be considered complete? Phys. Rev. **48**, 696–702 (1935)
8. M. Born, Zur Quantenmechanik der Stoßvorgänge. Z. für Phys. **37**, 863–867 (1926)
9. T. Breuer, Von Neumann, Gödel and quantum incompleteness, in *John von Neumann and the Foundations of Quantum Physics*, ed. by M. Rédei, M. Stöltzner (Kluwer, Dordrecht, 2001), pp. 75–82
10. L. de Broglie, La nouvelle dynamique des quanta, *Electrons et Photons: Rapports et Discussions du Cinquième Conseil de Physique* (Gauthier-Villars, Paris, 1928), pp. 105–132
11. H. Brown, G. Svetlichny, Nonlocality and Gleason's lemma. Part I. Deterministic theories. Found. Phys. **20**, 1379–1387 (1990)
12. H. Brown, C.G. Timpson, Bell on Bell's theorem: the changing face of nonlocality (2014), arXiv:1501.03521
13. J. Bub, Is von Neumann's 'no hidden variables' proof silly? in *Deep Beauty: Mathematical Innovation and the Search for Underlying Intelligibility in the Quantum World*, ed. by H. Halvorson (Cambridge University Press, Cambridge, 2011), pp. 393–408
14. J. Butterfield, Bell's theorem: what it takes. Br. J. Philos. Sci. **43**, 41–83 (1992)
15. C.S. Calude, *Information and Randomness: an Algorithmic Perspective*, 2nd edn. (Springer, Berlin, 2002)
16. C.S. Calude, Algorithmic randomness, quantum physics, and incompleteness. Lect. Notes Comput. Sci. **3354**, 1–17 (2004)
17. E. Cator, N.P. Landsman, Constraints on determinism: Bell versus Conway–Kochen. Found. Phys. **44**, 781–791 (2014)

18. R. Clifton, Getting contextual and nonlocal elements-of-reality the easy way. Am. J. Phys. **61**, 443–447 (1993)
19. G.J. Chaitin, A theory of program size formally identical to information theory. J. Assoc. Comput. Mach. **22**, 329–340 (1975)
20. G.J. Chaitin, *Information, Randomness and Incompleteness: Papers on Algorithmic Information Theory* (World Scientific, Singapore, 1987)
21. J.H. Conway, S. Kochen, The strong free will theorem. Not. Am. Math. Soc. **56**, 226–232 (2009)
22. L. Corry, *David Hilbert and the Axiomatization of Physics (1898–1918): from Grundlagen der Geometrie to Grundlagen der Physik* (Kluwer, Dordrecht, 2004)
23. E. Crull, G. Bacciagaluppi, *Grete Hermann-Between Physics and Philosophy* (Springer, Berlin, 2016)
24. A. Dasgupta, Mathematical foundations of randomness, in *Handbook of the Philosophy of Science. Volume 7: Philosophy of Statistics*, ed. by P.S. Bandyopadhyay, M.R. Forster (North-Holland/Elsevier, Amsterdam, 2011), pp. 641–710
25. P. Diaconis, B. Skyrms, *Ten Great Ideas About Chance* (Princeton University Press, Princeton, 2018)
26. D. Dieks, Von Neumann's impossibility proof: mathematics in the service of rhetorics. Stud. Hist. Philos. Mod. Phys. **60**, 136–148 (2016)
27. R. Downey, D.R. Hirschfeldt, *Algorithmic Randomness and Complexity* (Springer, Berlin, 2010)
28. R. Downey, D.R. Hirschfeldt, A. Nies, S.A. Terwijn, Calibrating randomness. Bull. Symb. Log. **12**, 411–491 (2006)
29. D. Dürr, S. Goldstein, N. Zanghi, Quantum equilibrium and the origin of absolute uncertainty. J. Stat. Phys. **67**, 843–907 (1992)
30. A. Eagle, Chance versus randomness, *The Stanford Encyclopedia of Philosophy* (2019), https://plato.stanford.edu/archives/spr2019/entries/chance-randomness/
31. J. Earman, *A Primer on Determinism* (D. Reidel, Dordrecht, 1986)
32. J. Earman, Curie's principle and spontaneous symmetry breaking. Int. Stud. Philos. Sci. **18**, 173–198 (2004)
33. A. Einstein, B. Podolsky, N. Rosen, Can quantum-mechanical description of physical reality be considered complete? Phys. Rev. **47**, 777–780 (1935)
34. M. Esfeld, Bell's theorem and the issue of determinism and indeterminism. Found. Phys. **45**, 471–482 (2015)
35. T. Franzén, *Gödel's Theorem: an Incomplete Guide to Its Use and Abuse* (AK Peters, Natick, 2005)
36. P. Gács, Review of "algorithmic information theory" by Gregory J. Chaitin. J. Symb. Log. **54**, 624–627 (1989)
37. K. Gödel, Über formal unentscheidbare Sätze der Principia Mathematica und verwandter Systeme I. Monatshefte für Mathematik und Physik **38**, 173–198 (1931)
38. S. Goldstein, Bohmian mechanics, *The Stanford Encyclopedia of Philosophy* (2017), https://plato.stanford.edu/archives/sum2017/entries/qm-bohm/
39. G. Greenstein, *Quantum Strangeness: Wrestling with Bell's Theorem and the Ultimate Nature of Reality* (The MIT Press, Cambridge, 2019)
40. P.D. Grünwald, P.M.B. Vitányi, Algorithmic information theory, in *Handbook of the Philosophy of Information*, ed. by P. Adriaans, J. van Benthem (Elsevier, Amsterdam, 2008), pp. 281–320
41. R. Haag, *Local Quantum Physics: Fields, Particles, Algebras* (Springer, Berlin, 1992)
42. I. Hacking, *The Taming of Chance* (Cambridge University Press, Cambridge, 1990)
43. B. Hensen et al., Experimental loophole-free violation of a Bell inequality using entangled electron spins separated by 1.3 km. Nature **526**, 682–686 (2015)
44. P. Heywood, M. Redhead, Nonlocality and the Kochen–Specker paradox. Found. Phys. **13**, 481–499 (1983)
45. D. Hilbert, Mathematical problems. Lecture delivered before the international congress of mathematicians at Paris in 1900. Bull. Am. Math. Soc. **8**, 437–479. Translated from Göttinger Nachrichten **1900**, 253–297 (1902)

46. G. 't Hooft, *The Cellular Automaton Interpretation of Quantum Mechanics* (Springer Open, 2016), https://www.springer.com/gp/book/9783319412849

47. M. Kline, *Mathematics: The Loss of Certainty* (Oxford University Press, Oxford, 1980)

48. S. Kochen, E. Specker, The problem of hidden variables in quantum mechanics. J. Math. Mech. **17**, 59–87 (1967)

49. A.N. Kolmogorov, *Grundbegriffe de Wahrscheinlichkeitsrechnung* (Springer, Berlin, 1933)

50. M. van Lambalgen, Random sequences. Ph.D. thesis, University of Amsterdam (1987), https://www.academia.edu/23899015/RANDOM_SEQUENCES

51. M. van Lambalgen, Randomness and foundations of probability: Von Mises' axiomatisation of random sequences, *Statistics, Probability and Game Theory: Papers in Honour of David Blackwell*. IMS Lecture Notes–Monograph Series, vol. 30 (1996), pp. 347–367

52. K. Landsman, *Foundations of Quantum Theory: from Classical Concepts to Operator Algebras* (Springer Open, 2017), https://www.springer.com/gp/book/9783319517766

53. K. Landsman, Randomness? What randomness? Found. Phys. **50**, 61–104 (2020)

54. K. Landsman, Quantum theory and functional analysis, in *Oxford Handbook of the History of Interpretations and Foundations of Quantum Mechanics*, ed. by O. Freire, to appear (Oxford University Press, Oxford, 2021)

55. L.A. Levin, On the notion of a random sequence. Sov. Math.-Dokl. **14**, 1413–1416 (1973)

56. M. Li, P.M.B. Vitányi, *An Introduction to Kolmogorov Complexity and Its Applications*, 3rd edn. (Springer, Berlin, 2008)

57. J. Mehra, H. Rechenberg, *The Historical Development of Quantum Theory. Vol. 6: the Completion of Quantum Mechanics 1926–1941. Part 1* (Springer, Berlin, 2000)

58. R. von Mises, Grundlagen der Wahrscheinlichkeitsrechnung. Math. Z. **5**, 52–99 (1919)

59. R. Montague, Deterministic theories, in *Formal Philosophy: Selected Papers of Richard Montague*, ed. by R.H. Thomason (Yale University Press, London, 1974), pp. 303–360

60. J.R. Moschovakis, Iterated definability, lawless sequences and Brouwer's continuum, in *Gödel's Disjunction*, ed. by L. Horsten, P. Welch (Oxford University Press, Oxford, 2016), pp. 92–107

61. A.A. Muchnik, A.L. Semenov, V.A. Uspensky, Mathematical metaphysics of randomness. Theor. Comput. Sci. **207**, 263–317 (1998)

62. J. von Neumann, *Mathematische Grundlagen der Quantenmechanik* (Springer). English translation: *Mathematical Foundations of Quantum Mechanics* (Princeton University Press, Princeton, 1955) (1932)

63. A. Nies, *Computability and Randomness* (Oxford University Press, Oxford, 2009)

64. T. Norsen, Local causality and completeness: Bell vs Jarrett. Found. Phys. **39**, 273–294 (2009)

65. M.B. Pour-El, J.I. Richards, *Computability in Analysis and Physics* (Cambridge University Press, Cambridge, 2016)

66. P. Raatikainen, On interpreting Chaitin's incompleteness theorem. J. Philos. Log. **27**, 569–586 (1998)

67. M. Rédei, M. Stöltzner (eds.), *John von Neumann and the Foundations of Quantum Physics* (Kluwer, Dordrecht, 2001)

68. M. Redhead, *Incompleteness, Nonlocality, and Realism* (Oxford University Press, Oxford, 1989)

69. J. Renn (ed.), *The Genesis of General Relativity*, vol. 4 (Springer, Berlin, 2007)

70. M. Scriven, The present status of determinism in physics. J. Philos. **54**, 727–741 (1957)

71. M.P. Seevinck, J. Uffink, Not throwing out the baby with the bathwater: Bell's condition of local causality mathematically 'sharp and clean'. Explan. Predict. Confirmation **2**, 425–450 (2011)

72. I. Sen, A. Valentini, Superdeterministic hidden-variables models I: nonequilibrium and signalling (2020), arXiv:2003.11989, II: conspiracy, arXiv:2003.12195

73. G. Senno, A computer-theoretic outlook on foundations of quantum information. Ph.D. thesis, Universidad de Buenos Aires (2017)

74. W. Sieg, *Hilbert's Programs and Beyond* (Oxford University Press, Oxford, 2013)

75. P. Smith, *An Introduction to Gödel's Theorems* (Cambridge University Press, Cambridge, 2013)

76. A. Stairs, Quantum logic, realism, and value definiteness. Philos. Sci. **50**, 578–602 (1983)

77. K. Svozil, *Randomness and Undecidability in Physics* (World Scientific, Singapore, 1993)
78. K. Svozil, *Physical (A)Causality: Determinism, Randomness and Uncaused Events* (Springer Open, 2018), https://www.springer.com/gp/book/9783319708140
79. K. Svozil, C.S. Calude, M.A. Stay, From Heisenberg to Gödel via Chaitin. Int. J. Theor. Phys. **44**, 1053–1065 (2005)
80. J. Szangolies, Epistemic horizons and the foundations of quantum mechanics (2018), arXiv:1805.10668
81. K. Tadaki, A refinement of quantum mechanics by algorithmic randomness (2018), arXiv:1804.10174
82. K. Tadaki, *A Statistical Mechanical Interpretation of Algorithmic Information Theory* (Springer, Berlin, 2019)
83. W. Tait, *The Provenance of Pure Reason: Essays in the Philosophy of Mathematics and Its History* (Oxford University Press, Oxford, 2005)
84. C. Tapp, *An den Grenzen des Endlichen: Das Hilbertprogramm* (Springer, Berlin, 2013)
85. S.A. Terwijn, The mathematical foundations of randomness, in *The Challenge of Chance: a Multidisciplinary Approach from Science and the Humanities*, ed. by K. Landsman, E. van Wolde (Springer, Berlin, 2016), pp. 49–66, https://www.springer.com/gp/book/9783319262987
86. A. Valentini, Foundations of statistical mechanics and the status of the Born rule in de Broglie-Bohm pilot-wave theory (2019), arXiv:1906.10761
87. S.B. Volchan, What is a random sequence? Am. Math. Mon. **109**, 46–63 (2002)
88. R.F. Werner, M.M. Wolf, Bell inequalities and entanglement. Quantum Inf. Comput. **1**, 1–25 (2001)
89. H.M. Wiseman, The two Bell's theorems of John Bell. J. Phys. A **47**, 424001 (2014)
90. S. Wolf, Nonlocality without counterfactual reasoning. Phys. Rev. A **92**, 052102 (2015)
91. N.S. Yanofsky, *The Outer Limits of Reason* (The MIT Press, Cambridge, 2013)
92. R. Zach, Hilbert's finitism: historical, philosophical, and metamathematical perspectives, Ph.D. thesis, University of California at Berkeley (2001)
93. A.K. Zvonkin, L.A. Levin, The complexity of finite objects and the development of the concepts of information and randomness by means of the theory of algorithms. Russ. Math. Surv. **25**, 83–124 (1970)

Chapter 4
Unpredictability and Randomness

Rade Vuckovac

Abstract Randomness is to a certain degree opposite to determinism. This essay tries to put those two on the same page. It argues the premise where randomness is a consequence of a deterministic process. It also provides yet another viewpoint on the hidden variable theory.

4.1 Introduction

Any discussion about randomness should include a description of unpredictability. A quoted part from the explanation on true randomness makes a good starting point for our inquiry (bold emphasis added):

> If outcomes can be determined (by hidden variables or whatever), then any experiment will have a result. More importantly, any experiment will have a result whether or not you choose to do that experiment, because the result is written into the hidden variables before the experiment is even done. Like the dice, if you know all the variables in advance, then you don't need to do the experiment (roll the dice, turn on the accelerator, etc.). The idea that every experiment has an outcome, regardless of whether or not you choose to do that experiment is called "the reality assumption", and it should make a lot of sense. If you flip a coin, but don't look at it, then it'll land either heads or tails (this is an unobserved result) and it doesn't make any difference if you look at it or not. In this case the hidden variable is "heads" or "tails", and it's only hidden because you haven't looked at it.

> It took a while, but hidden variable theory was eventually disproved by John Bell, who showed that there are lots of experiments that cannot have unmeasured results. Thus the results cannot be determined ahead of time, so there are no hidden variables, and the results are truly random. **That is, if it is physically and mathematically impossible to predict the results, then the results are truly, fundamentally random.** [1]

Some deterministic systems which show unpredictable behaviour are in:

R. Vuckovac (✉)
Department of Information and Communication Technologies, Engineering School,
Universitat Pompeu Fabra, Barcelona, Spain
e-mail: rade.vuckovac@gmail.com

Chaos Theory; It shows one exciting feature not usually found in classical systems. Predicting a long-term state of a system combined with its approximate initial conditions is a difficult if not impossible task. Edward Lorenz sums it as:

> Chaos: When the present determines the future, but the approximate present does not approximately determine the future.

Then following details show unpredictability in a little bit more detail:

- *Butterfly Effect;* An interesting and not widely emphasised point about Butterfly Effect is [2, 3]:

  > Even a tiny change, like the flap of a butterfly's wings, can making a big difference for the weather next Sunday. This is the butterfly effect as you have probably heard of it. But Edward Lorenz actually meant something much more radical when he spoke of the butterfly effect. He meant that for some non-linear systems you can only make predictions for a limited amount of time, even if you can measure the tiniest perturbations to arbitrary accuracy.

- *n-body problem;* Even in Newton times, the motions of more than two orbiting bodies were considered as a problem (Fig. 4.1). Currently, we are left with numerical methods and simulations. The former is an approximation and butterfly effect prone. The later are basically computational experiments, which are the preferable option for *n-body* system investigation [4, 5].

Cellular Automation (CA); CA is a deterministic model containing grids, populated by cells. The grids are arranged in one or two-dimensional space. An evolution rule governs how the initial state of cells evolve to the next generation. Figure 4.2 shows one dimensional CA rules and an evolution history. One of the interesting CA features is the concept of Computational Irreducibility (CI). It proposes that the only way to determine the future state of CA is to run it, which is very similar to "the results cannot be determined ahead of time" principle in hidden variable argumentation.

4.2   CA a Closer Look

While Chaos Theory provides us with dynamical systems showing unpredictability behaviour, the CI (computational irreducibility) principle is probably more accessible for the discussion.

Wolfram's rule 30 CA is a good starting point. Figure 4.2 shows the transition rules on the top. Every case prescribes how a cell (black or white) is transformed depending on the previous cell state and its neighbouring cells. Row 1 is the initial state of CA. Rows 2, 3, 4 ... are consecutive evolved generations. A next-generation cell is derived from a previous cell and its neighbours. For example, the cell (row 4, column 13) is derived from case 7. When a cell does not have an above row neighbour, the cell from other end is considered. So, for example, the cell (16; 1) uses case 7 (again) because the top left neighbour is the cell (15; 31).

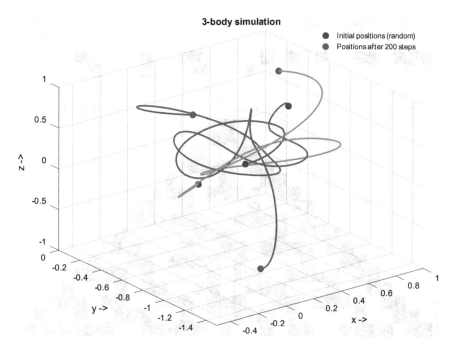

3-body simulation

● Initial positions (random)
● Positions after 200 steps

Fig. 4.1 An example of chaotic behaviour when the system has more than two bodies

There are a contest and prize for solving three CA rule 30 problems [6]. All three of them are relevant for the randomness discussion:

Problem 1: *Does the centre column always remain non-periodic?* The centre column is column 16 outlined red in Fig. 4.2. The period has been checked for the $\approx$ first billion steps, and the centre column does not show periodicity. In effect, it has very similar properties to π.

Problem 2: *Does each colour of cell occur on average equally often in the centre column?* The ratio of black and white cells approaches 1:1 when the step iterations increase. After one billion steps, the centre column has 500025038 black and 499974962 white cells. In a sense, it mimics a fair coin flipping exercise.

Problem 3: *Does computing the nth cell of the centre column require at least $O(n)$ computational effort?* Stephen Wolfram strongly suspects that rule 30 is computationally irreducible (CI). In that case, even if the initial state and rules of transformation are known, the quickest way to see the future of the state is to run CA (to do the experiment).

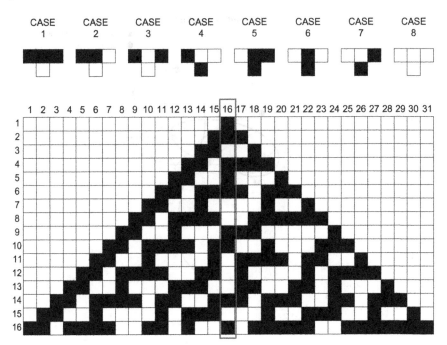

Fig. 4.2 CA rule 30. Transformation rules and evolution history. Column 16 acts as a random sequence

4.3 Conditional Branching

While chaos theory and CA provide evidence of future inaccessibility, there is an even more persuasive argument where we can not acquire the future state of some systems without experimenting.

Conditional branching, the underlying algorithmic construct is the essential argumentation ingredient. Conditional branching with other two primitives, sequence and iteration, provides all necessary blocks to build any algorithm imaginable [7]. It is an if-else statement in a program. When the program reaches if-else command, it evaluates some state and depending on evaluation continues with an if or an else path. Usually, when the program is made, all the possible inputs and they eventual execution paths (EP) are thoroughly defined. Figure 4.4 shows domain partition, where inputs are partitioned according to the joint EP.

For example, the $3x + 1$ problem one step is:

$$F(x) = \begin{cases} f(x) \ if & x \equiv 0 \bmod 2 \\ g(x) \ if & x \equiv 1 \bmod 1 \end{cases} \tag{4.1}$$

where $f(x) = x/2$, and $g(x) = (3x + 1)/2$. There we know that even input will be executed by $f(x)$ and odd with $g(x)$.

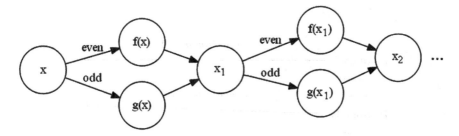

Fig. 4.3 The $3x + 1$ composite function; $f(x) = x/2$ and $g(x) = (3x + 1)/2$

Table 4.1 For two iterations of $3x + 1$ (Fig. 4.3), we have four partitions of natural numbers depending on execution path and corresponding composite

Domain ($\mathbb{N}_1$)	Execution path (EP)	Composite function
Partition A	Even-even	$f(f(x))$
Partition B	Even-odd	$g(f(x))$
Partition C	Odd-odd	$g(g(x))$
Partition D	Odd-even	$f(g(x))$

The problem starts when this procedure is iterated (Fig. 4.3 and Table 4.1). Every iteration doubles the amount of unique EP. For example if our inputs are 64-*bit* integers and we iterate Eq. 4.1 64 times. The number of possible paths is $EP \leq |2^{64}|$. In that case domain partition by execution paths (Fig. 4.4) seems an impossible task and that is probably the cause of why this problem is still not solved despite dealing with very basic arithmetic [8].

Now we can ask: Is every program with multiple execution paths domain partition-able?

Informal Theorem 1. *There exists at least one algorithm with multiple execution paths where the knowledge of which way execution goes is known only after input evaluation* [9].

Proof We can assume the opposite: every input partitioning can be performed efficiently before execution. On first glance, that is a reasonable statement because when a program is specified correctly, every case behaviour is fully defined in advance. On the other hand, there are at least two problems with this:

• Programs are not always made with a purpose. For example, multiple branching statements could be written haphazardly throughout a program. If it compiles, it will run, but the output behaviour will be unpredictable.
• If every program is partition-able, then branching algorithmic structure is redundant. Practically, we will know which path to execute for some input in advance without the need to test the branching statement.

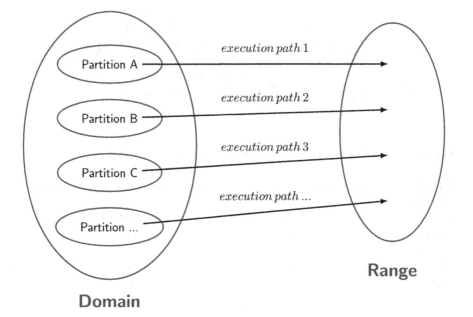

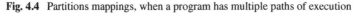

Fig. 4.4 Partitions mappings, when a program has multiple paths of execution

Having algorithms without conditional branching is not what we presently believe:

– Encyclopedia Britannica; Conditional branching entry:

> In Analytical Engine ⋯ control transfer, also known as conditional branching, whereby it would be able to jump to a different instruction depending on the value of some data. This extremely powerful feature was missing in many of the early computers of the 20th century. [10]

– Wikipedia; Turing Completeness:

> To show that something is Turing complete, it is enough to show that it can be used to simulate some Turing complete system. For example, an imperative language is Turing complete if it has conditional branching. [11]   □

4.3.1 Conditional Branching Candidates

While the systems where non-partitionable inputs exist, it is hard to identify one. Some candidates are:

CA rule 30; English description of the rule is in the exact form as Eq. 4.1 (bold added):

> Look at each cell and its right-hand neighbor. **If** both of these where white on the previous step, then take the new color of the cell to be whatever the previous color of its left-hand neighbor was. **Otherwise**, take the new color to be opposite of that. [12]

Rule 30 used to be a random number generator in Wolfram's Mathematica software.

Collatz conjecture; Equation 4.1 is one step of the procedure. The conjecture asserts that every natural number after some steps reaches 1. Some reflections on the problem:

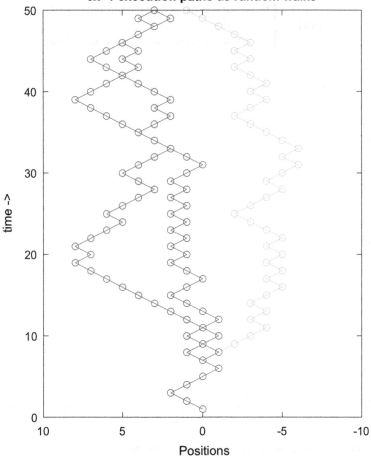

Fig. 4.5 Random walk using $3x + 1$ execution paths. When the algorithm executes the odd branch $((3x + 1)/2)$ the left step is taken; otherwise, the walk takes the right step $(x/2)$

The iterates of the shift function are completely unpredictable in the ergodic theory sense. Given a random starting point, predicting the parity of the n-th iterate for any n is a "coin flip" random variable. ... Empirical evidence seems to indicate that the 3x + 1 function on the domain Z retains the "pseudorandomness" property on its initial iterates until the iterates enter a periodic orbit. This supports the 3x + 1 conjecture and at the same time deprives us of any obvious mechanism to prove it, since mathematical arguments exploit the existence of structure, rather than its absence. [13, p. 18]

Figure 4.5 shows random walks behaviour using branching 3x + 1 parity.

Möbius function; It is another mathematical object with a branching structure and randomness emergence (Fig. 4.6). It appears that the Riemann hypothesis, random walks and Möbius function are closely related. A Math StackExchange discussion on this topic is here [14]. Equation 4.2 and Fig. 4.6 show the Möbius function, its branching structure and a random behaviour emergence.

$$\mu(n) = \begin{cases} 0 & \text{if } 0 \text{ if n has a squared prime factor} \\ 1 & \text{if } n = 1 \\ (-1)^k & \text{if n is a product of k distinct primes} \end{cases} \tag{4.2}$$

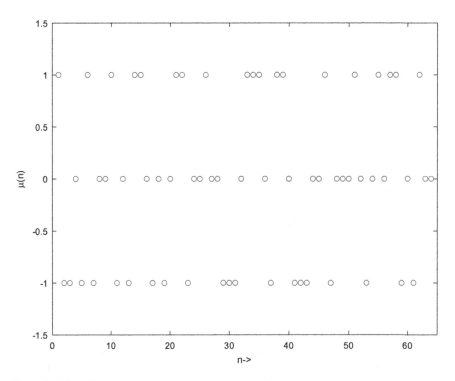

Fig. 4.6 Möbius function and its randomly looking mappings

Branching CA (BCA); This CA was at first designed as a toy model to investigate the if-else structure and its impact on the state transitions. The CA evolution step has now a familiar structure:

$$c' = \begin{cases} c \oplus A_{i+1}, & \text{if } A_{i+2} > A_{i+3} \\ c \oplus \overline{A}_{i+1}, & \text{otherwise} \end{cases} \tag{4.3}$$

where $\oplus$ is exclusive or and $\overline{A}_{i+1}$ is one complement of the cell A_{i+1}. This particular CA is used as a cryptographic primitive for cipher design [15]. Figure 4.7 using *branching CA* illustrates all chaos theory features including the butterfly effect.

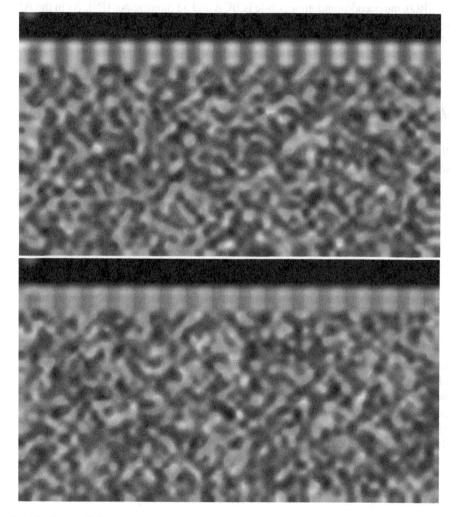

Fig. 4.7 Two cellular automata evolutions where the initial state (top left three pixels plus top black region) differs by just one bit (one in 4096-*bit* state)

4.4   Measuring Randomness

While all mentioned candidates show randomness, the measure of it is a seemingly
tricky endeavour. The Turing test for intelligence [16] may provide the base for
randomness evaluation. A similar idea was explored in cryptography as well [17].

We have three entities in Turing AI test: a human, a machine and an interrogator.
They are in three separate rooms, and their communication is in the questions/answer
form where interrogator asks, and the other two parties respond. The evaluation of
answers should determine who the human is. If a distinction is inconclusive, the
machine could be considered intelligent.

For randomness game we have true-randomness side represented by Random Oracle (RO), the pseudo-randomness side is BCA and an interrogator (IN). As in the AI
test, IN communicate with RO and BCA via questions/answer (input/output) format.
IN task is to decide which party produces truly random output. If the distinction
could not be made, the true and pseudo qualifiers are redundant.

4.4.1   Random Oracle

RO concept provides a specific tabular representation which we can categorise as a
true random mapping:

The following model describes a random oracle:

- There is a black box. In the box lives a gnome, with a big book and some dice.
- We can input some data into the box (an arbitrary sequence of bits).
- Given some input that he did not see beforehand, the gnome uses his dice to generate
  a new output, uniformly and randomly, in some conventional space (the space of oracle
  outputs). The gnome also writes down the input and the newly generated output in his
  book.
- If given an already seen input, the gnome uses his book to recover the output he returned
  the last time, and returns it again. [18]

The RO model is believed to be an imaginary construct useful in a cryptography
secure protocol argumentation. However, it cannot be realised in practice. The use
of dices in the description is assumed to be an entirely random process. In other
words, if we need RO in practice, we have to use a random function. Its main feature
is that every output is an independent, random string. That implies the use of a
gigantic input/output table (Table 4.2). That table cannot be compressed, rendering
the exercise impractical. More details about RO can be found here [19].

In the finding pseudo game, IN (interrogator) asks a question and RO (the gnome)
replies with a truly random answer with the stipulation that same question is answered
with the same response.

Table 4.2 Random function

Domain (N)	Range (N)
Input 1	Random string 1
Input 2	Random string 2
Input 3	Random string 3
...	...

Fig. 4.8 BCA transformation rule; depending on the neighbours relation, third right-hand neighbour or its one complement (all bits are flipped) is XORed (logical exclusive or) with the carry. Updated carry is XORed with the current cell to form a new cell

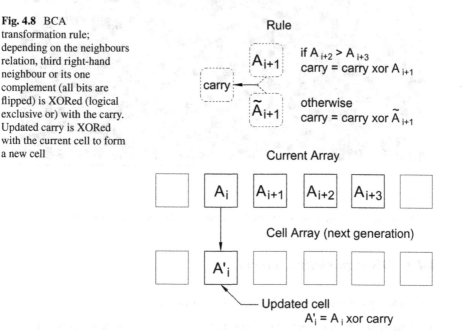

4.4.2   BCA

BCA candidate (mentioned earlier) represents a pseudo-random oracle. Transformation details are shown in Fig. 4.8. When IN (interrogator) submits the question the following happens: firstly an array is initialised. The input x (question) is copied on the array first cells. The array is evolved for some time, and the part of the last state serves as output y (response). Figure 4.9 red and blue are input and output used in IN and BCA communications.

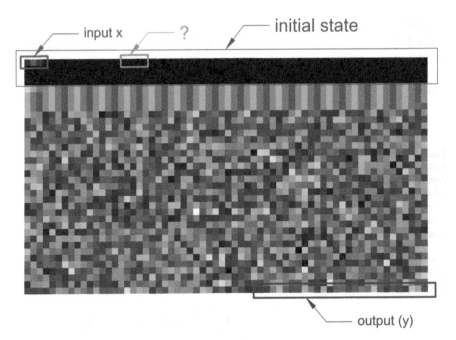

Fig. 4.9 Branching cellular automata (BCA) pretending to be a random oracle (RO)

4.4.3 Distinguishing Chances

The finding pseudo can be played by two set of rules:

- 1st **variant**: IN (interrogator) knows inputs and corresponding outputs only (red and blue Fig. 4.9).

  When IN examines BCA input/output pairs, a couple of exciting things are revealed. Similar inputs, produce very different outputs (Fig. 4.7). When numerous input/output pairs are put in a tabular form, the table starts to reassemble random structure from random function mappings (Table 4.2). This set-up is reminiscent to the Alan Turing challenge scenario:

  > I have set up on the Manchester computer a small programme using only 1,000 units of storage, whereby the machine supplied with one sixteen-figure number replies with another within two seconds. I would defy anyone to learn from these replies sufficient about the programme to be able to predict any replies to untried values. [16]

- 2nd **variant**; we give IN some hints: IN can have full knowledge of BCA algorithm and only partial knowledge of the initial state. Meaning, knowing the magenta part excluding the green part Fig. 4.9. If IN wants to check whether the observed pair comes from BCA by running it, the green part has to be replaced with some provisional part. To find a matching green part IN has to perform an exhaustive search. The same happens if the IN goes backward from the output. Complete knowledge of the state is needed (provisional plus output) to be able to proceed with steeping. That situation leaves IN with brute force option only.

In both variants, the IN (interrogator) inability to correlate input/output pairs stays. The 2nd variant also prevents IN to check if observed data comes from BCA even the inner working of the input/output transformation is known.[1]

4.5   Speculations

The main point raised in this essay is the computation barrier imposed by conditional branching. That wall forces input evaluation during program execution. Even with known input and algorithm knowledge, predictability of output is impossible. The only way to discover output is to run the algorithm (to evaluate input). The absence of patterns, when input/output pairs of such algorithm were analysed, is caused by that barrier.

If we assume that the conditional barrier is relevant to physical reality and in the context of:

> Thus the results cannot be determined ahead of time, so there are no hidden variables, and the results are truly random.

We can speculate; yes, the results can not be determined ahead of time but, paradoxically, *that does not necessary exclude determinism.*

The misunderstandings might come from determinism definitions because we have different meanings which usually are not noted in argumentations. There are at least two sets of determinism with very other properties. The differentiation occurs when we need to access state details at a particular point of time. For example, imagine, a special kind of video player which can play two movies only: 2-*body* digital simulation and the sequel *n-body* digital simulation. While watching the 2-*body* movie, we can fast forward it if we are bored. While watching *n-body* movie, we notice that fast forward is not working any more, and if we want to see what happens on end, we have to watch the whole movie.

From there, we can identify potential inconsistencies when dealing with determinism. In the accepted hidden variable theories (HVT) narrative, a cursory look at the Bell's inequality [20] and HVT indicates that we have determinism acting locally with the correspondent distribution of probable outcomes which is never confirmed by experiment. Since quantum distribution is matched with observed, we conclude that quantum phenomena are non-deterministic.

On the other hand, we have determinism from the *n-body* movie. Desired state details are known only when they happen. Any imaginable distribution information

---

[1] The partially knowing BCA initial state case might be analogous to the true randomness cause. Generally, initial state ignorance comes from practical reasons. Fidelities of real state versus assumed state might be tiny, and misreadings can be easily made. We can see what one-bit difference in 4096-*bit* state can do for BCA evolution (Fig. 4.7). There is a good chance that RO (true) and BCA (pseudo) randomness have the same origin.

is obtained by observation only. It is not evident how this kind of determinism fits in HVT context and how its distributions can be violated by experiment.

For the end, there is a parallel with apparent quantum weirdness in the algorithmic world:

$3x + 1$ *problem story*; We have a state (a natural number). The state will go through the transition, as shown in Fig. 4.1. Before the measurement, the state is in a superposition of a number of possible execution paths. Only after running the algorithm (measurement), the superposition state collapses, and we know the execution path and the transition result.

The $3x + 1$ problem is still unsolved. The conjecture support comes from experimental evidence where all numbers between 1 and $2^{\approx 59}$ are reaching 1 as postulated. Another support comes from a heuristic probability which shows that every multiplication (with 3) has two divisions (with 2) on average, which indicates a starting number shrinkage on the long run [8]. Both of the supports or what we know about the phenomena comes from experiments and probability distributions; even it is a deterministic process.

References

1. Ask a Physicist. Do physicists really believe in true randomness? https://www.askamathematician.com/2009/12/q-do-physicists-really-believe-in-true-randomness/
2. S. Hossenfelder, The 10 most important physics effects, http://backreaction.blogspot.com/2020/02/the-10-most-important-physics-effects.html
3. E.N. Lorenz, The predictability of a flow which possesses many scales of motion. Tellus **21**(3), 289–307 (1969)
4. K.T. Alligood, T.D. Sauer, J.A. Yorke, Chaos. Springer (1996)
5. M. Trenti, P. Hut, N-body simulations (gravitational). Scholarpedia **3**(5), 3930 (2008)
6. The wolfram rule 30 prizes, https://www.rule30prize.org/
7. C. Böhm, G. Jacopini, Flow diagrams, turing machines and languages with only two formation rules. Commun. ACM **9**(5), 366–371 (1966)
8. Collatz conjecture, https://en.wikipedia.org/wiki/Collatz_conjecture
9. R. Vuckovac, On function description (2020), arXiv:2003.05269
10. Conditional branching, https://www.britannica.com/technology/conditional-branching
11. Turing completeness, https://en.wikipedia.org/wiki/Turing_completeness
12. S. Wolfram, A new kind of science (rule 30 alternative description p 27), https://www.wolframscience.com/nks/p27--how-do-simple-programs-behave/?firstview=1/
13. J.C. Lagarias, The 3x + 1 problem: an overview, http://bookstore.ams.org/mbk-78/FreeAttachments/mbk-78-prev.pdf
14. Riemann hypothesis, random walks and möbius function, https://math.stackexchange.com/questions/2350052/riemann-hypothesis-random-walks-and-m
15. R. Vuckovac, Secure and computationally efficient cryptographic primitive based on cellular automaton. Complex Syst. **28**(4), 457–474 (2019)
16. A.M. Turing, Computing machinery and intelligence (1950), in *The Essential Turing: the Ideas That Gave Birth to the Computer Age*, ed. by B.J. Copeland (Oxford University Press, Oxford, 2004), pp. 433–464
17. O. Goldreich, Randomness, interactive proofs, and zero-knowledge–a survey, *The Universal Turing Machine, A Half Century Survey* (Citeseer, 1988)

18. What is the "random oracle model" and why is it controversial? https://crypto.stackexchange.com/questions/879/what-is-the-random-oracle-model-and-why-is-it-controversial
19. What is the random oracle model and why should you care? https://blog.cryptographyengineering.com/2011/09/29/what-is-random-oracle-model-and-why-3/
20. J.S. Bell, On the Einstein Podolsky Rosen paradox. Phys. Phys. Fiz. **1**(3), 195 (1964)

Chapter 5
Indeterminism, Causality and Information: Has Physics Ever Been Deterministic?

Flavio Del Santo

Abstract A tradition handed down among physicists maintains that classical physics is a perfectly deterministic theory capable of predicting the future with absolute certainty, independently of any interpretations. It also tells that it was quantum mechanics that introduced fundamental indeterminacy into physics. We show that there exist alternative stories to be told in which classical mechanics, too, can be interpreted as a fundamentally indeterministic theory. On the one hand, this leaves room for the many possibilities of an open future, yet, on the other, it brings into classical physics some of the conceptual issues typical of quantum mechanics, such as the measurement problem. We discuss here some of the issues of an alternative, indeterministic classical physics and their relation to the theory of information and the notion of causality.

5.1 When Did Physics Become Unpredictable?

Like all of the human activities, also science maintains traditions that are handed down from generation to generation and help to form the identity of a community. One specific story that seems to have crystallized among practitioners is that classical physics (i.e., Newton's mechanics and Maxwell's electrodynamics) would allow, in principle, to predict everything with certainty. The standard story continues by telling that the foundations of such a theory are perfectly well understood and free of any interpretational issues. In particular, it is widely accepted that classical physics categorically entails a *deterministic* worldview.

Indeed, due to the tremendous predictive success of Newtonian physics (in particular in celestial mechanics), it became customary to conceive an in principle limitless predictability of the physical phenomena that would faithfully reflect the fact that

F. Del Santo (✉)
Institute for Quantum Optics and Quantum Information (IQOQI-Vienna), 1090 Vienna, Austria
e-mail: delsantoflavio@gmail.com

Faculty of Physics, University of Vienna, 1090 Vienna, Austria

Basic Research Community for Physics (BRCP), Vienna, Austria

© The Author(s), under exclusive license to Springer Nature Switzerland AG 2021
A. Aguirre et al. (eds.), *Undecidability, Uncomputability, and Unpredictability*,
The Frontiers Collection, https://doi.org/10.1007/978-3-030-70354-7_5

our Universe is governed by determinism. This view was advocated and vastly pop-
ularized in the early nineteenth century by Pierre-Simon Laplace, who envisaged the
possibility for a hypothetical superior intelligence –which went down in history as
Laplace's demon– to predict the future states of the universe with infinite precision,
given a sufficient knowledge of the laws of nature and the initial conditions [1]:

> Given for one instant an intelligence which could comprehend all the forces by which nature
> is animated and the respective situation of the beings which compose it –an intelligence
> sufficiently vast to submit these data to analysis– it would embrace in the same formula the
> movements of the greatest bodies in the universe and those of the lightest atom; to it nothing
> would be uncertain, and the future as the past would be present to its eyes.

The standard story goes on by stating that this faith in perfect determinism was
abruptly shattered by the advent of quantum theory, which, with its probabilistic
predictions, made indeterministic doubts burst into physics for the first time.[1] But is
it really so? In this essay, we will show that this is not necessarily the case and that
the alleged fundamental difference between classical and quantum physics based on
their alleged inherently deterministic, respectively indeterministic, character should
be rethought.

From the historical point of view, as early as 1895 –thus before that any quantum
effect was discovered and such theory formulated– the father of the kinetic theory
of gases, Ludwig Boltzmann, already doubted the very possibility of having perfect
determinism at the microscopic scale (see [2, 3]). It ought to be noticed that according to the standard understanding of classical statistical mechanics, probabilities are
there introduced to account for a lack of knowledge about the actual state of affairs
(epistemic randomness) and are not supposed to be *irreducible*.[2] This is due to the
fact that statistical physics deals with an enormous amount of components (and of
degrees of freedom), but it is generally accepted that every single classical particle
has a perfectly predetermined behavior (and that in principle this is predictable).
Despite this, Boltzmann maintained: "I will mention the possibility that the fundamental equations for the motion of individual molecules will turn out to be only
approximate formulas which give average values." [5]. Also Franz S. Exner, another
eminent Viennese physicists, contemporary of Boltzmann, questioned the validity
of determinism in classical physics even at the macroscopic level: "In the region of
the small, in time and space, the physical laws are probably invalid; the stone falls to
earth and we know exactly the law by which it moves. Whether this law holds, however, for each arbitrarily small fraction of the motion [...] that is more than doubtful."
[6]. One of the intellectual heirs of Boltzmann and Exner in Vienna was the Nobel
laureate and founding father of quantum theory, Erwin Schrödinger. Indeed, he fully
embraced their skeptical positions about determinism: "As pupil of the venerable
Franz Exner I have been on intimate terms for a long time with the idea that probably

---

[1] By indeterminism we denote the sufficient condition that there exists at least one phenomenon, or
a type of phenomena, which does not obey determinism.

[2] Probabilities are said to be irreducible if "it is not possible by further investigation to discover
further facts that will provide a better estimate of the probability." [4].

not microscopic lawfulness but perhaps 'absolute accident' forms the foundation of our statistics." [7].[3]

Interestingly, contrarily to the text-books presentation of classical physics, the fact that classical systems have a perfectly predeterminate dynamics (thus giving rise to perfectly deterministic predictions) is not inherent in the formalism (see Sect. 5.2). Rather, it is based on an additional hidden assumption that takes the form of a principle. In a previous work [9], we have named this *principle of infinite precision*. This is articulated in two parts, as follows:

> **Principle of Infinite Precision**
> 1. *Ontological*—there exists an actual value of every physical quantity, with its infinite determined digits (in any arbitrary numerical base).
> 2. *Epistemological*—despite it might not be possible to know all the digits of a physical quantity (through measurements), it is possible to know an arbitrarily large number of digits.

It is only when its formalism is complemented with this principle that classical physics becomes deterministic.

However, the principle of infinite precision is inconsistent with any operational meaning, as already made evident by Max Born. The latter gave pivotal contributions to the foundations of quantum formalism –introducing the fundamental rule that bears his name, which allows to assign probabilities to quantum measurements, and for which he was awarded the Nobel Prize in 1954– and became critical of determinism, even in classical physics, due to its reliance on "infinite precision". Indeed, in his essay *Is Classical Mechanics in fact Deterministic?* [10], he affirmed:

> It is usually asserted in this theory [classical physics] that the result is in principle determinate and that the introduction of statistical considerations is necessitated only by our ignorance of the exact initial state of a large number of molecules. I have long thought the first part of this assertion to be extremely suspect. [...] Statements like 'a quantity x has a completely definite value' (expressed by a real number and represented by a point in the mathematical continuum) seem to me to have no physical meaning. [Because they] cannot in principle be observed.

To explain how infinite precision and determinism relate to one another it is interesting to rephrase a simple example devised by Born. Referring to Fig. 5.1, consider a (classical) particle that is bound to move in a one-dimensional cavity with perfectly elastic walls and total length l. If the particle has a perfectly determinate (i.e. with infinite precision) initial position $x(t = 0) = x_0$ and velocity $\vec{v}(t = 0) = \vec{v}_0$, classical physics then allows to predict, also with infinite precision, the future positions $x(t)$ and velocities $\vec{v}(t)$ for any time instant t. Yet, imagine to slightly relax the principle of infinite precision and, while x_0 remains fully determinate, the initial

---

[3] This trend of doubting determinism in the Vienna school of statistical physics has been referred to as the *Vienna indeterminism* in the philosophical literature [8].

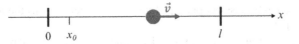

Fig. 5.1 Toy example of a system (a classical particle confined in a cavity), in which the indeterminacy on the initial conditions is amplified in the future indeterminacy of the physical state while time passes

velocity (say pointing to the right) of the particle has a small indeterminacy, i.e. $v_0 \leq v(t = 0) = v_0 + \Delta v_0$. Once the particle starts moving, its initial indeterminacy starts to be reflected on the determination of its position at later times. According to the laws of classical mechanics, the range of possible future positions of the particle increases linearly as time passes, i.e. $\Delta x(t) = t\Delta v_0$. This means that, for any arbitrarily small indeterminacy of the initial velocity Δv_0, there always exists a critical time instant $t_c := l/\Delta v_0$, such that $\Delta x(t = t_c) = l$. Namely, independently of how small is the initial indeterminacy, it is sufficient to wait enough long time for having complete indeterminacy on the particle's position within the cavity. This clearly shows that the principle of infinite precision is a necessary condition for determinism.

This example is one of the simplest instantiations of those systems whose future dynamics is highly susceptible to a variation of the initial conditions (a property called *instability*). Such a phenomenon is typical of the so-called *chaotic systems* wherein the uncertainty in the determination of future values of some physical quantities increases exponentially with elapsed time. Actually, our example displays what in chaos theory is referred to as the *real butterfly effect* [11]. Namely, not only high sensitivity to the initial conditions, but –since the system is bounded, and, accordingly, it is the region of phase space within which its physical state can evolve– after a certain time the uncertainty saturates the whole allowed region of phase space. This means that after a certain critical time the distribution of the states in phase space is the homogeneous one.[4] As a matter of fact, criticisms of classical determinism became more severe in the second half of the last century, when the chaos theory was further developed and its fundamental consequences understood [12, 13].

Furthermore, the challenges to determinism in classical physics experienced a revival in very recent years, when several scholars formalized the fact that predetermined physical quantities seems to be at odds with information-theoretic arguments [9, 14–19], as we will show in detail in what follows.

---

[4] We acknowledge Sabine Hossenfelder's essay "Math Matters" in the 2020 FQXi Essay Contest for Ref. [11].

5.2 The "Orthodox Interpretation" of Classical Physics

In order to introduce the arguments against the tenability of determinism in classical physics and a possible alternative interpretation thereof,[5] we first ought to recall some pillars of the formalism of that theory. We refer here to the standard formalism together with its metaphysical assumptions (i.e. the principle of infinite precision) as the "orthodox interpretation" of classical physics.[6]

Conceptually, classical physics (say Newtonian mechanics, but equivalent arguments apply to classical electromagnetism, too) is characterized by (i) the *physical state* of a system, which accounts for its relevant physical properties and (ii) a set of *general laws* that govern the evolution (backwards and forward in time) of the physical state.[7] Formally speaking, the dynamical properties of a system are identified by a set of physical quantities, which mathematically are called variables. The collection of these variables (typically position and momentum) is called the *state* of the system and the space composed of all the possible values taken by these quantities is called *phase space*. Moreover, the assumption of the principle of infinite precision results in the fact that classical states are mathematical points in a continuous phase space. Namely, a physical state is an n-tuple of real numbers for a n-dimensional phase space.

As for the mathematical characterization of the general laws of classical mechanics, these are ordinary differential equations that take as inputs the values from the state at time t_0, called *initial conditions*, and return the state of the system at any arbitrary time t. The mathematical theory of differential equation then guarantees that, given any sequence of subsequent states in a certain interval of time, i.e. a *trajectory* in phase space for that interval, there exists always a unique extension thereof, into the past and the future [21]. This mathematical formalization all together leads, in fact, to the formal definition of *Laplacian determinism*: For any given physical state there exists a unique evolution, i.e., a unique trajectory in phase space. However, as stressed by Drossel, "the idea of a deterministic time evolution represented by a trajectory in phase space can only be upheld within the framework of classical mechanics if a point in phase space has infinite precision" [18].

We have already pointed out how the concept of infinite precision has no operational meaning. This was also recently remarked by Rovelli, who stated that "concretely we never determine a point in phase space with infinite precision –this would be meaningless–, we rather say that the system 'is in a finite region R of phase space', implying that determining the value of the variables will yield values in R." [22].

---

[5] We use here the expression "interpretation" and not "theory" because we consider only empirically indistinguishable predictions (in the same sense of the interpretations of quantum mechanics) [20].

[6] We borrow this name from the foundations of quantum mechanics, where the attribution "orthodox" is usually associated to the most widespread interpretation of the quantum formalism, also called the "Copenhagen interpretation", attributed to Niels Bohr and his school.

[7] To these two main aspects of classical physics one has to add a third one, (iii) that there exists a background time which allows to speak about the state of a physical system at a certain instant of time and its evolution at later instants.

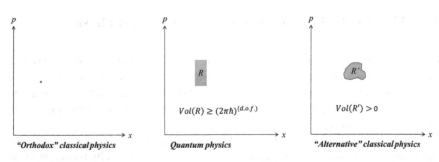

Fig. 5.2 Suggestive representations of physical states in phase space, in comparison for "orthodox" classical physics (left), quantum physics (middle), and "alternative" classical physics (right). In the orthodox classical physics, the state is a mathematical point which determines a unique trajectory (determinism), whereas both in quantum and alternative classical physics the state has a fundamental indeterminacy that leads to an indeterministic dynamics. The peculiarity of quantum physics is that its formalism sets a precise value to the smallest size of a cell in phase space

Note that, however, as most of his fellow physicists, Rovelli upholds the "orthodox" classical mechanics which considers this a for-all-practical-purposes issue, i.e. at the fundamental level, R degenerates to a mathematical point. In quantum physics, on the other hand, there is a fundamental lower limit to the size of the region in phase space. In particular, the volume of the region, $Vol(R)$, cannot be smaller than the size delimited by the Planck constant (for each degree of freedom), i.e.,

$$Vol(R) \geq (2\pi \hbar)^{(d.o.f.)}.$$

And Rovelli refers to this as "*the* major physical characterization of quantum theory" [22].

But are quantum and classical physics necessarily so different on this matters? One has to realize that the principle of infinite precision is not part of the mathematical formalism of classical theory, but rather it belongs to the domain of interpretations. In fact, one can consider –and several arguments point in the direction that one should perhaps do so– an alternative interpretation of classical physics in which physical states are not mathematical points characterized by (n-tuples of) real numbers. In this way, even classical physics would display a fundamental indeterminacy, and its conceptual difference with quantum mechanics should be scaled down (see Fig. 5.2).

Invoking again Born's operationalism, one ought to consider the following [10]:

A statement like $x = \pi$ cm would have a physical meaning only if one could distinguish between it and $x = \pi_n$ cm for every n, where π_n is the approximation of π by the first n decimals. This, however, is impossible; and even if we suppose that the accuracy of measurement will be increased in the future, n can always be chosen so large that no experimental distinction is possible. Of course, I do not intend to banish from physics the idea of a real number. It is indispensable for the application of analysis. What I mean is that a physical situation must be described by means of real numbers in such a way that the natural uncertainty in all observations is taken into account.

As we will show in the next section, one can indeed envision an alternative classical physics that maintains the same general laws (equations of motion) of the standard formalism, but dismisses the physical relevance of real numbers, thereby assigning a fundamental indeterminacy to the values of physical quantities, as wished by Born. In fact, "as soon as one realizes that the mathematical real numbers are "not really real", i.e. have no physical significance, then one concludes that classical physics is not deterministic." [15].

5.3 An Alternative, Indeterministic Interpretation of Classical Physics

5.3.1 Determinism at Odds with Information Principles

The relaxation of the principle of infinite precision does not come about only as a mere intellectual exercise, or as a proof of principle that classical physics is compatible with alternative interpretations beyond the orthodox one. In fact, the motivation for searching novel interpretations of classical physics stems also from the application of information-theoretic concepts to physics. Indeed, our current understanding tells us that

> information is not a disembodied abstract entity; it is always tied to a physical representation. It is represented by engraving on a stone tablet, a spin, a charge, a hole in a punched card, a mark on paper, or some other equivalent. This ties the handling of information to all the possibilities and restrictions of our real physical world, its laws of physics and its storehouse of available parts. [23].

This view goes under the name of *Landauer's principle*, in short, "information is physical".

In Ref. [15], Gisin gave sound arguments to support the claim that "a finite volume of space cannot contain more than a finite amount of information". Intuitively, this is a direct consequence of Landauer's principle, because each bit of information to be stored requires a certain amount of space, bounded from below by the size of the smallest physical system that can encode it. It is true that today, thanks to the incredible development of the technology of miniaturization, we are able to encode and manipulate information in astonishingly small systems. This allows to reach densities of information storage of about 25 terabytes per centimeter square [24] on atomic lattices, whereas molecular storage of information in DNA has recently achieved extraordinary densities of information of the order of a million terabyte per cubic millimeter [25]. These outstanding results notwithstanding, physical systems have a finite size, hence it seems a very reasonable assumption to believe that there is a finite limit to the possible information density.

Furthermore, a well known formal theoretical argument sets a limit to the allowed information density, called *Bekenstein bound* [26], states that the information I (in number of bits) contained in a system circumscribed by a sphere of radius R is smaller than the mass-energy E enclosed in the same sphere, i.e.,

$$\frac{I}{2\pi R} \lesssim \frac{E}{\ln 2},$$

where we have adopted the Planck units (i.e., $c = \hbar = 1$). The intuition behind this is that the storage of each bit of information is associated with a certain amount of energy and that unbound densities of energy degenerate into black holes.

Coming back to the orthodox interpretation of classical physics, we have already shown how this assumes that physical states are mathematical points in phase space, expressed by (n-tuples of) real numbers. However, it should be noticed that real numbers contain, in general, an infinite amount of information. As we have learnt since primary school, the set of real numbers encompasses all the familiar rational numbers and supplement them with the irrational numbers. However, even among the irrational numbers, there are fundamental conceptual differences that have relevant consequences for the role they are attributed in physics. All the irrational numbers we are used to speak about, such as $\sqrt{2}$ or π, are, in fact, *computable* irrational numbers. This means that they can be compressed into an algorithm of finite length which, at every iteration, outputs the next digit of the considered number. For instance, an algorithm (but not the only one) to construct π is given by computing (each digit of) the ratio of the circumference of any circle to its diameter. So, although an irrational computable number has infinite digits without a periodic pattern, and, as such, it would take infinite time (i.e., iterations of the associated algorithm) to get all the digits, its actual information content is finite. Everything there is to know about it is contained in the algorithm that generates it. More precisely, the (finite) information content of a computable number corresponds to the amount of information in bits of shortest algorithm that outputs that number (i.e., its *Kolmogorov complexity*). What is however disconcerting, is that the amount of computable numbers among all the real numbers is infinitely small (i.e., it forms a subset of Lebesgue measure zero). Technically, the probability of picking a computable number from the set of real numbers is zero (see also [14, 15]).

Putting together the above arguments, we come to the conclusion that real numbers cannot be physically meaningful insofar as their information content is almost always infinite. One thus ought to consider alternative interpretations of classical physics that do not enforce the principle of infinite precision. Namely, interpretations that do not assume that physical quantities take values in the real numbers. Note again that without real numbers, one cannot any longer uphold determinism in classical physics.

In this view, the orthodox interpretation of classical physics can be regarded as a deterministic completion of an indeterministic model, in terms of *hidden variables*: Namely, the real numbers [15, 16]. This is reminiscent of Bohm's [27] or Gudder's [28] hidden variable models of quantum physics, which provide a deterministic description of quantum mechanics by adding (in principle inaccessible) supplementary variables, whereas the orthodox interpretation takes probabilities (therefore indeterminism) to be irreducible.

5.3.2 "Finite Information Quantities" (FIQs)

To overcome the problem of the infinite information content of real numbers in the context of physics, an explicit alternative model has been sketched in Ref. [15] and developed in greater detail in Ref. [9]. This model entails an alternative indeterministic interpretation of classical mechanics. We review here its main features.

In the spirit of the previous considerations, let us leave the dynamical equations of Newtonian mechanics unchanged, but let us relax the principle of infinite precision by substituting the real numbers with newly defined quantities. We refer to them as "finite-information quantities" (FIQs), which, while providing the same empirical predictions as the orthodox interpretation of classical physics, have no overlap with real numbers (they are not a mathematical number field, nor a proper subset thereof).

Let us start by considering again the orthodox interpretation. Let a physical quantity $\gamma \in \mathbb{R}$ (say the position of a particle moving in one dimension) lie, without loss of generality, in the interval $[0, 1]$ and write it in binary base:

$$\gamma = 0.\gamma_1\gamma_2 \ldots \gamma_j \ldots,$$

where $\gamma_j \in \{0, 1\}$, $\forall j \in \mathbb{N}^+$. This means that, being $\gamma \in \mathbb{R}$, its infinite bits are *all* given at once, i.e., always determined.

Consider now the following alternative model to describe physical quantities which introduces an element of randomness in such a way to always guarantee the finiteness of the information content. We thus introduce the following:

> **Definition—*propensities*** There exist objective properties, named *propensities*, $q_j \in [0, 1] \cap \mathbb{Q}$, for each digit j of a physical quantity. A propensity quantifies the tendency of the jth binary digit to take the value 1.

The concept of propensities, borrowed from Popper's objective interpretation of probabilities [29], can be understood from the limit cases, namely when they are either 0 or 1. For example, $q_j = 1$ means that the jth digit will take value 1 with certainty. On the opposite end, if a bit has an associated propensity of 1/2, it means that the bit is totally indeterminate. We posited that propensities are rational numbers, but in general it is enough that their information content is always finite (e.g., they could be computable real numbers). In order to define physical quantities, we thus have to define the following:

> **Definition—*FIQs*** A *finite-information quantity* (FIQ) is an ordered list of propensities $\{q_1, q_2, \cdots, q_j, \cdots\}$, each associated to a bit of a physical quantity, such that the overall information content is finite, i.e., $\sum_j I_j < \infty$, where I_j is the information content of the jth propensity (as expressed by some reasonable measure).

Note that a previous work [9], we have suggested a straightforward way to construct a FIQ, i.e. to assume that after a certain threshold, all the bits to which propensities are associated become completely random, i.e., $\exists M(t) \in \mathbb{N}$ such that $q_j = 1/2, \forall j > M(t)$. In this way, the propensities are all independent and it is possible to choose $I_j = 1 - H(q_j)$, where H is the binary entropy function of its argument. It's trivial to check that in that scenario $\sum_j I_j < \infty$. However, in Ref. [30], it was pointed out a weakness of the simple latter scenario, namely that the mutual independence between propensities of a FIQ is not preserved under a basic operation such as a change of unit. We have shown in [31] that this does not jeopardizes the FIQ program, but indeed forces us to introduce more complex ways to construct FIQs, such that correlations between propensities are properly introduced. Despite the criticism in [30], for conceptual simplicity we still present in what follows the simple model to construct FIQs using independent propensities, because this will more intuitively allow to discuss the main conceptual novelties and issues.

In general, since we require our alternative interpretation to be empirically equivalent to the orthodox one, at least the digits of a physical variable that are already known (i.e., measured) at time t should be fully determined. Therefore, the propensities of the first, more significant, $N(t)$ digits should be already actualize, i.e. $q_i \in \{0, 1\}, \forall i \leq N(t)$. We are now ready to express a physical quantities γ in this FIQ-based interpretation:

$$\gamma\,(N(t), M(t)) = 0.\ \underbrace{\gamma_1 \gamma_2 \ldots \gamma_{N(t)}}_{\text{determined } \gamma_j \in \{0,1\}}\ \overbrace{?_{N(t)+1} \ldots ?_{M(t)}}^{?_k,\ \text{with } q_k \in (0,1)}\ \underbrace{?_{M(t)+1} \cdots}_{?_l,\ \text{with } q_l = \frac{1}{2}},$$

where the symbols $?_i$ means that the digit in position i is not yet determined. Notice that in this framework the potential property of becoming actual (a list of propensities, FIQ), has somehow a more fundamental status than an already actualized value (a list of determined bits). In fact, in this alternative interpretation, a state would be the collection of all the FIQs associated with the dynamical variables (i.e., the list of the propensities of each digit). Thus, even two systems that are to be considered identical at a certain instant of time (in the sense that they are in the same state) will have, in general, different actual values at later times. But then, how does the actualization happen in such a way that is compatible with the observed results? To answer this question we need to discuss what is a measurement in a non-deterministic physics.

5.3.3 The Classical "Measurement Problem"

Any indeterministic interpretation of a physical theory needs to face the questions (*measurement problem*): How does a single value of a physical variable become actualized out of its possible values? Or how does potentiality become actuality? Our experience, in fact, tell us that every time a quantity gets measured there is only one value registered by the instrument. In order to address this issue, however, it is necessary to first ask: What is a measurement? This long-lasting question is one of the most profound open problems of the foundations of quantum physics (see, e.g. [32]). As we have recalled, the latter is normally considered the first theory to have introduced fundamental indeterminacy in the domain of physics. Yet, as soon as an indeterministic interpretation of classical physics is upheld, this is also subject to a measurement problem.

Let us operationally define what are the minimal requirement for a process to be considered a measurement:

> **Definition—*Minimal requirements for a measurement***
> 1. *Stability*: Consecutive measurements of the same quantity leave the already determined digits unchanged.
> 2. *Intersubjectivity*: Different agents can access the same measurement outcomes.
> 3. *Precision improvability*: With more accurate measurement apparatuses, more digits become available (with the former two properties).

As for stability, it should be remarked that it is of course possible that the dynamical evolution would change the state of the system under consideration, and therefore the outcomes of measurements occurring at two (arbitrarily distant) instants of time. However, what is assumed here is a trivial evolution, or equivalently a short enough time interval between consecutive measurements, thus focusing only on the changes of the states due to measurements. This assumption is customarily upheld in explaining quantum mechanics, when one says that two consecutive measurements of the same observable yield the same results (obviously if performed in the same basis).

In order to be an empirically adequate model, in the FIQ-based indeterministic interpretation of classical physics, too, one needs to explain how to comply with these properties of measurements. By construction of FIQs, propensities are objective properties subjected to fundamental irreversibility (i.e., once they become either 0 or 1 the remain unchanged), and this accounts for the stability and ensures the intersubjective availability of the measurement results.

What is far from being straightforward, however, is the compliance with precision improvability. In Ref. [9], we have introduced two possible ways to account for this property. On the one hand, one can envisage (i) a mechanism that makes the actualization to spontaneously occur as time passes. This resembles the so-called *objective collapse models* of quantum mechanics [33–35]. On the other hand, it is possible to think that (ii) the actualization happens when a higher level requires it, thus, with some *top-down causation* mechanism [36]. In this case, it would be the measurement apparatus that "imposes" to the physical variables to acquire a determined value. This is clearly reminiscent of the Copenhagen interpretation of quantum mechanics.

Such a "classical measurement problem" remains an open problem as much as its more notorious quantum counterpart. Yet, it should be noticed that however problematic, the fact that both classical and quantum physics share this issue helps to scale down the fundamental difference between these two theories.

5.4   (In)determinism and Causality

Like in the orthodox interpretation, in the indeterministic model previously introduced, too, the laws of classical mechanics are taken to be general relations that causally connect physical states at different instants of time. Traditionally, in the philosophy of science, the concepts of determinism and causality have been long wedded, to the extent that usually Laplacian determinism is often referred to as *causal determinism*. One of the most notable examples of this can be found in Hume, who maintained that a cause is always sufficient for its effect: "It is not possible on Hume's account, for causes to be less than deterministic." [4]. Also Leibniz elevated determinism to an a priori truth, when formulating his *principle of sufficient reason*: "There is nothing without a reason, or no effect without a cause" (quoted in [21]). And Kant even formulated what is sometimes called the *law of universal causation*, according to which, "if we [...] experience that something happens, then we always presuppose thereby that something precedes on which it follows in accordance with a rule." [37].[8]

The concept of causation is also traditionally related, at least in science, to the quest for explanation. This means to ask: Why did the (observed) event E_i happen? (here i labels the time at which E occurs). Answering this question in a deterministic worldview seems to us quite meaningless. In fact, were *everything* completely predetermined, this question –like any other one– would not be a genuine question, in the sense that, for whoever asked it, this was a necessity from the beginning of times. Recalling again Laplace's demon, "the future as the past would be present to its eyes" [1]. In other words, everything would just be an already shot film that is unrolling, and the script of the film makes you say the line "why did E_i happen?".

Moreover, even if we assume that an agent, or an intellect, is external to *everything* that occurs in the universe (i.e. s/he is really watching the movie from outside), thereby not being included in this predetermined state of the universe, this is not unproblematic. Asking *why* something happened is in this case certainly meaningful but the answer is trivial: Because this is the film I am watching (and there are no

---

[8] Note that Kant refers to the term *rule* as a univocal correspondence and does not contemplate any non-deterministic (e.g., probabilistic) law that relates causes and effects.

other films available!). Again, determinism assumes that given an initial state of the universe and universal laws everything *causally* follows. But this is misleading because there is only one specific initial state and, without alternatives, causation seems a void concept.

On the contrary, indeterminism introduced the possibility of alternatives, thereby making causality meaningful. If one asks the reason why a certain event E_j occurred, is now possible to reply: "Because another event $E_{i/A}$ happened before (i.e., $i < j$) and not its mutually exclusive alternative $E_{i/B}$". Significant progress in weakening the bond between determinism and causality was made in the second half of the nineteenth century, thanks to the work of philosophers the likes of Popper [29], Earman [21], Salmon [38], Dowe [4], Reichenbach [39], Good [40] and Suppes [41]. Mostly inspired by quantum mechanics, the concept of *probabilistic causality* came about. This maintains that an event C directly influences another event E but is not sufficient for it. A common, and quite grim, example to explain probabilistic causality features the following chain of events (temporally ordered): A scientist, Eric, sits in a sealed room (i.e., without any exchange with the external environment). His colleague, Clara, brings a canister full of radioactive material in Eric's room (ideally, making sure that there are no other exchanges with the environment). While time elapses, the radioactive material will be decaying –at a certain probabilistic rate depending on its chemical composition– releasing ionizing radiation. Sadly, at some point, Eric develops radiation poisoning. Now, since decay is governed by quantum mechanics and in that theory probabilities are considered irreducible, there was no deterministic process relating Clara's actions to Eric's condition. However, if you think that Clara can be held accountable for Eric's sickness, then you believe in probabilistic causality.

Referring to Fig. 5.3, we can graphically formalize deterministic (on the left) and probabilistic causality (on the right). Both are represented as graphs which are *directed* (causes precede their effects in time), and *acyclic* (an effect cannot be the cause if itself). However, in a deterministic graph, there are no possible alternatives: Everything that can happen does happen.[9] On the contrary, a graph representing probabilistic causality (Fig. 5.3-right) is a *multigraph* with two types of edges. The first ones (blue) represent the "potential causations" and are *weighted* with the measure or the degree to which an event E_i causes future events $E_{j/K}$, where i, j label the time instants and $K \in \{A, B, \ldots\}$ the possible mutually exclusive alternatives.[10] A natural choice for the weights of the potential causation is clearly propensities q_K, as defined in Sect. 5.3.2. The second kind of edges (red), instead, represent what actually happened and can be reconstructed in hindsight after the actualization of the potentiality has happened (e.g., after measurements).

---

[9] Note that it is of course not necessary that each event is effect and cause of one and only one event as in Fig. 5.3-left. We represented this simple chain because we deem less confusing the comparison with the probabilistic graph (Fig. 5.3-right).

[10] Note that the celebrated *Many-World Interpretation* of quantum mechanics affirms that all the possible alternative outcomes actually happen, hence refuting the mutual exclusiveness thereof. While this is also a possible further interpretation of the FIQ-based physics, we will not consider this further.

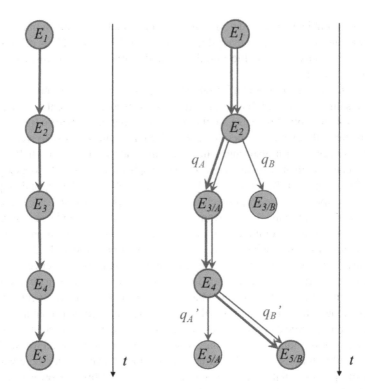

Fig. 5.3 Causal graphs of a deterministic (left) and indeterministic (right) series of events. The red arrows relate events that actually happen, while the blue ones denote potential, mutually exclusive, alternative events of which only one will happen with a certain propensity q_K (see main text)

Clearly, the alternative interpretation of classical physics based on FIQs, introduced in Sect. 5.3.2, is causal but not deterministic and can be represented by causal graphs of the second type (Fig. 5.3-right).

Recently, D'Ariano, Manessi and Perinotti [42] have pointed out that the notions of determinism and causality are logically independent, namely one can have not only non-deterministic (probabilistic) causal theories, such as quantum mechanics, but in principle non-causal deterministic theories, too. Their argument is carried out in the framework of *operational probabilistic theories*, introduced by some of the same authors in previous works. Without entering the formal details, according to the authors of Ref. [42] a theory is said to be causal if there is *no-signaling from the future*. Namely, if the probability of preparing a system in a certain initial state is independent from the choice of measurements that will be performed on the system itself. Determinism is instead defined as "the property of a theory of having all probabilities of physical events equal to either zero or one." [42]. They then cleverly design a toy-theory that is, in fact, deterministic but not causal, according to their definitions.

However, defining deterministic behaviors as a limiting case of probabilistic ones, while is seemingly very natural, leads to subtle issues. Indeed, this boils down to give an interpretation of what probabilities are supposed to mean. If they are taken to have no causal meaning, but being merely measured frequencies of occurrences, then the fact that determinism and causality are logically independent becomes trivial. Consider a typical example of classical correlations: During a vacation in New York, take a pair of shoes and separate them into two identical boxes. Shuffle the boxes in a way that is impossible to know which one contains the left, respectively the right, shoe. Keep one box with you and send the other to a friend in Tokyo. You can now open the box and you figure out that you kept, say, the left shoe. Then you can infer the following statement with probability one, i.e., deterministically: "My friend in Tokyo has received a right shoe". Nobody, however, would ever entail that finding the left shoe in New York has caused the right show to be found in Tokyo.[11]

Similarly, take the digits of a (computable) number, say π. If you know with certainty that the number you are dealing with is really π, for instance because is the ratio between a circle's circumference and its diameter, then you can assert with probability one, i.e. deterministically, that its nineteenth decimal digit is a 4. Again, nobody would claim that you caused this digit to be 4 by measuring the circumference and diameter of a circle.

5.5 Concluding Remarks

In this essay, we have revised arguments to support the view that classical physics could be interpreted indeterministically, and basic operational principles and information-theoretic arguments hint at this direction. At the same time, quantum mechanics has given us reason (in particular by means of the violation of *Bell's inequalities* [43]) to believe that the universe we live in is not deterministic. If this is the case, Popper's words remind us of what is the reward for indeterminism: "The future is open. It is not predetermined and thus cannot be predicted –except by accident. The possibilities that lie in the future are infinite" [44]. In fact, not only is the future unpredictable in an indeterministic universe, but also the truth values of future (scientific) statements are genuinely *undecidable*, as Gisin's simple example points out [45]:

> Think of a proposition about the future, for example, "It will be raining in exactly one year time from now at Piccadilly Circus". If one believes in determinism, then this proposition is either true or false [...]. But if one believes that the future is open, then it is not predetermined that it will rain, hence the proposition is not true, and it is not predetermined that it will not rain, thus the proposition is also not false.

---

[11] One can object that in fact, the events were causally determined by the operation of shuffling and it is only subjective ignorance that makes this appear random. Fair enough, but then substitute the shoes with two quantum entangled particles and you will convince yourself that you have "determinism" (in the sense of perfect correlations) without causality (see [43]).

However, our stance on an open future cannot remain but a belief, because compelling arguments (e.g., [46, 47]) show that every physical theory, including classical and quantum mechanics, can be interpreted either deterministically or indeterministically and no experiment will ultimately discriminate between these two opposite worldviews. We can only have the certainty that the future of the battle between determinism and indeterminism is open, too.

Acknowledgements I would like to show my gratitude to Nicolas Gisin and Borivoje Dakić for the many interesting discussions and their comments that helped to improve this essay. I am indebted also to the many, interesting discussions held on the forum of FQXi Community during the 2020 Essay Contest. I also acknowledge Marco Erba for pointing out Ref. [42] to me.

References

1. P.-S. Laplace, *A Philosophical Essay on Probabilities* (English translation by W.F. Truscott, F.L. Emory, Dover Publications, New York, 1951) (1820)
2. M. Jammer, Indeterminacy in physics, *Dictionary of the History of Ideas*, vol. 2 (Charles Scribner's Sons, New York, 1973), pp. 586–594
3. F. Del Santo, Striving for realism, not for determinism: historical misconceptions on Einstein and Bohm. APS News **18**(5) (2019)
4. P. Dowe, *Physical Causation: Cambridge Studies in Probability, Induction, and Decision Theory* (Cambridge University Press, Cambridge, 2000)
5. L. Boltzmann, *Vorlesungen Uber Gastheorie* (J.A. Barth, Leipzig, 1896) (English translation by H.S.G. Bus, University of California Press and Cambridge University Press, Berkeley, 1964)
6. F. Exner, *Über Gesetze in Naturwissenschaften und Humanistik* (Vienna, 1909)
7. Schrödinger to Bohr, 24 May 1924, Reproduced in P.A. Hanle, Indeterminacy before Heisenberg: the case of Franz Exner and Erwin Schrödinger. Hist. Stud. Phys. Sci. **10**, 225–269 (1979)
8. M. Stöltzner, Vienna indeterminism: Mach, Boltzmann, Exner. Synthese **119**(1–2), 85–111 (1999)
9. F. Del Santo, N. Gisin, Physics without determinism: alternative interpretations of classical physics. Phys. Rev. A **100**(6), 062107 (2019)
10. M. Born, *Physics in My Generation* (Springer, New York, 1969)
11. T.N. Palmer, A. Doring, G. Seregin, The real butterfly effect. Nonlinearity **27**, 123 (2014)
12. D.S. Ornstein, Ergodic theory, randomness, and chaos. Science **243**(4888), 182–187 (1989)
13. I. Prigogine, I. Stengers, *The End of Certainty* (Simon and Schuster, New York, 1997)
14. G. Dowek, Real numbers, chaos, and the principle of a bounded density of information, in *International Computer Science Symposium in Russia* (Springer, Berlin, 2013)
15. N. Gisin, Indeterminism in physics, classical chaos and Bohmian mechanics. Are real numbers really real? Erkenntnis (2019). https://doi.org/10.1007/s10670-019-00165-8
16. N. Gisin, Real numbers as the hidden variables of classical mechanics, *Quantum Studies: Mathematics and Foundations*, in press (2019)
17. S.J. Blundell, Emergence, causation and storytelling: condensed matter physics and the limitations of the human mind. Philosophica **92**, 139–164 (2017)
18. B. Drossel, On the relation between the second law of thermodynamics and classical and quantum mechanics, *Why More Is Different* (Springer, Berlin, 2015)
19. P. Lynds, Time and classical and quantum mechanics: indeterminacy versus discontinuity. Found. Phys. Lett. **16**(4), 343–355 (2003)
20. V. Baumann, S. Wolf, On formalisms and interpretations. Quantum **2**, 99 (2018)

21. J. Earman, Determinism in the physical sciences, in *Introduction to the Philosophy of Science*, ed. by M.H. Salmon (Hackett, 1992)
22. C. Rovelli, Space is blue and birds fly through it. Philos. Trans. R. Soc. A: Math. Phys. Eng. Sci. **376**(2123), 20170312 (2018)
23. R. Landauer, The physical nature of information. Phys. Lett. A **217**(4–5), 188–193 (1996)
24. F. Kalff, M. Rebergen, E. Fahrenfort et al., A kilobyte rewritable atomic memory. Nat. Nanotechnol. **11**, 926–929 (2016)
25. L. Ceze, J. Nivala, K. Strauss, Molecular digital data storage using DNA. Nat. Rev. Genet. **20**, 456–466 (2019)
26. J. Bekenstein, Universal upper bound on the entropy-to-energy ratio for bounded systems. Phys. Rev. D. **23**(2), 287–298 (1981)
27. D. Bohm, A suggested interpretation of the quantum theory in terms of "hidden" variables. I. Phys. Rev. **85**(2), 166 (1952)
28. S.P. Gudder, On hidden-variable theories. J. Math. Phys. **11**(2), 431–436 (1970)
29. K.R. Popper, The propensity interpretation of probability. Br. J. Philos. Sci. **10**(37), 25–42 (1959)
30. L. Callegaro, F. Pennecchi, W. Bich, Comment on "physics without determinism: alternative interpretations of classical physics". Phys. Rev. A **102**(3), 036201 (2020)
31. F. Del Santo, N. Gisin, Reply to "comment on 'physics without determinism: alternative interpretations of classical physics'". Phys. Rev. A **102**(3), 036202 (2020)
32. Č. Brukner, On the quantum measurement problem, *Quantum [Un] Speakables II* (Springer, Berlin, 2017)
33. G.C. Ghirardi, A. Rimini, T. Weber, Unified dynamics for microscopic and macroscopic systems. Phys. Rev. D **34**(2), 470 (1986)
34. N. Gisin, Stochastic quantum dynamics and relativity. Helv. Phys. Acta **62**(4), 363–371 (1989)
35. G.C. Ghirardi, P. Pearle, A. Rimini, Markov processes in Hilbert space and continuous spontaneous localization of systems of identical particles. Phys. Rev. A **42**, 78 (1990)
36. G.F. Ellis, Top-down causation and emergence: some comments on mechanisms. Interface Focus **2**(1), 126–140 (2011)
37. I. Kant, *Critique of Pure Reason* (English translation by P. Guyer, A.W. Wood, Cambridge University Press, Cambridge, 1997) (1781)
38. W.C. Salmon, *Probabilistic Causality, in Causality and Explanation* (Oxford University Press, Oxford, 1998)
39. H. Reichenbach, *The Direction of Time* (University of California Press, Berkeley, 1956)
40. I.J. Good, A causal calculus. Br. J. Philos. Sci. **11**, 305–318 (1961)
41. P. Suppes, *A Probabilistic Theory of Causality* (North-Holland, Amsterdam, 1970)
42. G.M. D'Ariano, F. Manessi, P. Perinotti, Determinism without causality. Phys. Scr. **T163**, 014013 (2014)
43. J.S. Bell, On the Einstein Podolsky and Rosen paradox. Phys. Phys. Fiz. **1**(3), 195 (1964)
44. K.R. Popper, *The Myth of the Framework: in Defence of Science and Rationality* (Routledge, New York, 1994)
45. N. Gisin, Mathematical languages shape our understanding of time in physics. Nat. Phys. 1–3 (2020)
46. P. Suppes, The transcendental character of determinism. Midwest Stud. Philos. **18**(1), 242–257 (1993)
47. C. Werndl, Are deterministic descriptions and indeterministic descriptions observationally equivalent? Stud. Hist. Philos. Mod. Phys. **40**(3), 232–242 (2009)

Chapter 6
Undecidability, Fractal Geometry and the Unity of Physics

T. N. Palmer

Abstract An uncomputable class of geometric model is described and used as part of a possible framework for drawing together the three great but largely disparate theories of 20th Century physics: general relativity, quantum theory and chaos theory. This class of model derives from the fractal invariant sets of certain nonlinear deterministic dynamical systems. It is shown why such subsets of state-space can be considered formally uncomputable, in the same sense that the Halting Problem is undecidable. In this framework, undecidability is only manifest in propositions about the physical consistency of putative hypothetical states. By contrast, physical processes occurring in space-time continue to be represented computably. This dichotomy provides a non-conspiratorial approach to the violation of Statistical Independence in the Bell Theorem, one where key counterfactual states needed to establish Bell's theorem are undefined, thereby pointing to a possible causal deterministic description of quantum physics.

6.1 The Disunity of 20th Century Physics

Three of our greatest theories of physics were formulated in the 20th Century: general relativity theory, quantum theory and chaos theory. There is hardly any aspect of human endeavour in the 21st Century that has been untouched by the consequences of at least one of these theories. However, each is remarkably disparate from the others, the very antithesis of the unity to which most physicists aspire in their search for laws which govern the universe. To be specific:

A schematic of the local fractal state-space structure of the invariant set I_U in Invariant Set Theory. **a** An ensemble of trajectories decoheres into two distinct clusters labelled a and $a̸$. Under a second phase of decoherence, this trajectory, itself comprising a further ensemble, decoheres into two further distinct regions labelled b and $b̸$. **b** Under magnification, a trajectory segment is found to comprise a helix of p trajectories at the next fractal iterate. **c** Top: a cross section through the helix

T. N. Palmer (✉)
Department of Physics, University of Oxford, Oxford, UK
e-mail: tim.palmer@physics.ox.ac.uk

of trajectories comprises p (here $p = 16$) disks coloured black or grey according to whether that trajectory evolves to the a cluster or the $\cancel{a}$ cluster. Bottom: each of these p disks itself comprises p further disks coloured black or grey according to whether each trajectory evolves to the b or $\cancel{b}$. The fractal set C_p of disks is homeomorphic to the set of p-adic integers

- Our inability to synthesise general relativity theory and quantum theory into a satisfactory quantum theory of gravity is legendary and is widely regarded as the single biggest challenge in contemporary theoretical physics.
- There are profound differences between quantum theory and chaos theory despite the fact that unpredictability lies at the heart of both theories. In conventional interpretations of quantum theory, unpredictability arises from the randomness of the measurement process in what is otherwise a linear theory. By contrast, unpredictability arises in chaos theory from the instability and nonlinearity of its deterministic equations of motion. However, there is more than this. By virtue of its determinism, chaos has not been seen as a route to understand the phenomenon of quantum entanglement: in order to violate the Bell inequality a conventional chaotic model of quantum physics would have to be explicitly nonlocal, a property inimical to the goal of synthesising with a causal theory of gravity.
- The way chaos is typically defined is incompatible with the principles of relativistic invariance. In particular, a defining characteristic of a chaotic system is instability, characterised by the fact that two states which are initially close can diverge exponentially in time, implying the existence of positive so-called Lyapunov exponents [15]. However, such divergence can be eliminated by a logarithmic reparametrisation of time suggesting that, in terms of the standard definitions at least, the phenomenon of chaos is not coordinate independent [3].

The purpose of this essay is to provide some basis for believing that these theories can be brought closer together through the unifying concept of non-computability.

6.2 Chaos and the Undecidable Geometry of Fractal Attractors

Although unpredictability is a familiar if not defining characteristic of chaotic systems such as the famous three-component Lorenz equations [11]

$$\frac{dx}{dt} = \sigma(y - x)$$
$$\frac{dy}{dt} = x(\rho - z) - y$$
$$\frac{dz}{dt} = xy - \beta z, \tag{6.1}$$

chaotic unpredictability is not a direct manifestation of non-computability. The reason is as follows: although chaotic systems exhibit sensitive dependence to initial conditions, they do nevertheless exhibit continuous dependence on initial conditions [19]. Such continuous dependence means that in a chaotic system it is possible to predict reliably as far ahead as you like, providing the initial conditions are known sufficiently accurately. This implies that such predictions are computational (i.e. can be performed to arbitrary accuracy by finite computing machines in finite time).

However, in the infinite future, two interesting things happen. Firstly, no matter what the initial condition, the state of a time-irreversible chaotic system such as described by the Lorenz equations settles down on its fractal attractor, sometimes referred to as a dynamically invariant subset of state space, or invariant set for short. Secondly, in the infinite future, the property of continuous dependence on initial conditions finally breaks down. This suggests the interesting question: are the fractal attractors of chaotic dynamical systems uncomputable?

To answer this question, we must first define what is meant by the term 'uncomputable'. One can take the definition from the seminal work of Turing [21] who famously showed, by an extension of the Gödel incompleteness theorem, that no algorithm exists that can decide whether, from the set of all possible pairs of computer programs and program inputs, a given program-input pair will halt.

In their seminal book 'Complexity and Real Computation' [2] (co-authored by Steve Smale one of the pioneers of chaos theory), Blum et al. set about answering the question of whether membership of a fractal, such as the famous Mandelbrot Set, is decidable. Their argumentation applies equally to the fractal attractor $\mathcal{A}$ of a chaotic system. We consider a putative algorithm/machine on the real numbers, which takes as input a point $\mathbf{x}$ in the state space of the chaotic system, and halts if $\mathbf{x} \in \mathcal{A}$ (Fig. 6.1). Blum et al.'s Path Decomposition Theorem implies that such an algorithm does not exist if $\mathcal{A}$ does not have integer dimension. The very definition of a fractal is one whose (e.g. Hausdorff) dimension is not an integer. Hence we can conclude that indeed the fractal invariant sets of chaotic systems are uncomputable. This notion was further developed by Simant Dube [4], who showed that many of the classic undecidable problems of computing theory (e.g. the Post Correspondence Problem named after one of the pioneers of computing theory, Emile Post) can be recast in terms of geometric properties of fractal attractors (e.g. does a given line intersect the attractor).

It is possible that this property of non-computability may also arise in finite time, in the initial-value problem for the Navier-Stokes partial differential equations of classical fluid mechanics. The physics behind this assertion lies in the possibility that the e-folding time, associated with the linear instability of a particular turbulent eddy, decreases without bound as the spatial scale of the eddy goes to zero, implying a finite-time breakdown of continuous dependence on initial conditions [19]. However, such a property has not be proven rigorously and indeed is closely related to one of the Clay Mathematics Millenium Prize Problems. For this reason, we do not pursue it here. As the author has discussed in [19], this finite-time breakdown in the computability of the Navier Stokes equations (rather than unpredictability in low-order chaos) is what Ed Lorenz actually meant by 'The Butterfly Effect' [12].

Fig. 6.1 Adapted from Fig F
of [2]. Blum et al. consider a
putative algorithm/machine
that takes as input a point **x**
in the state space of a
dynamical system with
attractor $\mathcal{A}$, and halts if
$\mathbf{x} \in \mathcal{A}$. Their Path
Decomposition Theorem
implies that no such machine
exists if A has fractional
dimension. The fractal
attractors of chaotic systems
are therefore
non-computational, and
$\mathbf{x} \in \mathcal{A}$ is undecidable

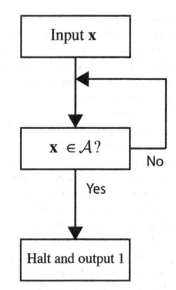

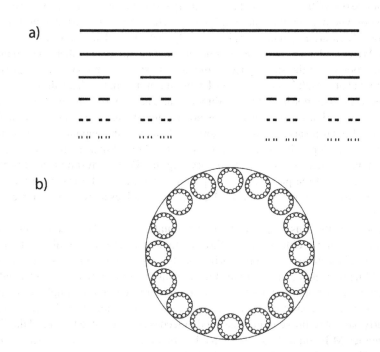

Fig. 6.2 Iterates of **a** the familiar Ternary Cantor Set $\mathcal{C}_2$; **b** a representation of the Cantor Set $\mathcal{C}_{16}$. In both cases, $\mathcal{C}_p$ is the intersection of all iterates. The set $\mathbb{Z}_p$ of p-adic integers is homeomorphic to $\mathcal{C}_p$

Before returning to the principal theme of this essay, a number of important points need to be made about properties of fractal attractors. The essence of a fractal attractor is the Cantor Set. In Fig. 6.2a we show a simple ternary Cantor set C_2 (remove the middle third from the interval [0, 1] and iterate). In Fig. 6.2b is shown a generalisation C_p to p iterated pieces (to which we return in the Appendix). In both cases, the fractal is itself the intersection of all fractal iterates. A point on C_2 can be represented by a base-3 real between 0 and 1 whose base-3 expansion does not contain the digit 1, e.g. $p = .02022020\ldots$. If we perturb this number by adding to it a number drawn from the unit interval (e.g. $\delta p = .0000010\ldots$), then almost certainly the perturbed number $p + \delta p$ will not lie on C_2.

This raises the question: How would we actually do mathematics on fractal attractors, for example so that when we add or multiply two points on such an attractor, the sum or product remains on the set? This is nontrivial because if, for example, we take two points $p_1 = .020220\ldots$ and $p_2 = .002020\ldots$ on C_2 (written in base 3) and add them together, then $p_1 + p_2 = .100010\ldots$ which contains the digits 1 and therefore does not lie on C_2. A similar issue arises if we multiply p_1 by p_2. One might imagine that simply replacing each 2 with a 1, so that p_1 and p_2 are represented in binary, would do the trick. However, it does not as the example $p_1 = .10010\ldots$, $p_2 = .11110\ldots$ shows (in this case $p_1 + p_2$ no longer lies in the unit interval [0, 1]).

In fact, there is a way to ensure that addition and multiplication on Cantor Sets are arithmetically closed. However, instead of using familiar real-number representations of p_1 and p_2 on C_2, we must instead use so-called 2-adic integer representations (and p-adic integers for C_p). p-adic numbers are bread and butter for pure mathematicians [9]. However, they are typically considered exotic by physicists. This may have to change if we are to exploit the notion of non-computability in physics. There is a rich theory of nonlinear dynamical systems based on mappings of p-adic numbers [23]. Going beyond this, it is possible to do calculus, complex analysis, Lie group theory, indeed much of the usual sorts of mathematics performed by physicists, using p-adic numbers. Like the real numbers, the set of p-adic numbers forms a completion of the set of rational numbers—however, with respect to a different metric: the p-adic metric rather than the more familiar Euclidean metric. There is an important consequence of this: points which lie in the fractal gaps of the Cantor Set (corresponding to p-adic numbers which are not p-adic integers) are, p-adically distant from points on the Cantor Set, even when from a Euclidean perspective they may seem arbitrarily close. This has conceptually implications discussed below.

Despite all this, the notion of fractals and non-computability may invoke a sense of uneasiness for physicists who believe that the world around us should be describable using finite mathematics. As Hilbert famously noted: 'the infinite is nowhere to be found in reality, no matter what experiences, observations, and knowledge are appealed to.' In this regard, we should note that non-computability can leave an imprint in finite approximations D' of chaotic systems D. In particular, the proposition $x \in A_{D'}$, although algorithmically decidable, can nevertheless be computationally irreducible: it cannot be decided reliably by an algorithm which is itself a simplification of D'. We will refer to this again in the discussion of the Bell Theorem.

In can be noted that finite representations of fractals can be represented simply by finite truncations of the corresponding p-adic integers.

6.3   Towards a Unification of 21st Century Physics

Using the concept and properties of uncomputable fractal attractors, let us return to the issue raised at the beginning of this essay: the disunity of the three great theories of 20th Century physics. We discuss possible ways to resolve the disunity in reverse order to that described above.

6.3.1   Chaos Theory and Relativity Theory

We can easily overcome the obstacle between chaos and relativity theory discussed above. The answer [3] is to define chaos in terms of the geometric properties of its fractal invariant set. For example, as discussed, one defining geometric characteristic of a fractal is its non-integer dimension. An approach based on an analysis of the invariant sets of a parametrisation of the cosmological Mixmaster model [13] allows one to talk meaningfully about coordinate-independent chaos in a relativistically invariant cosmological setting. This allows us to introduce a concept which is central to the discussion of quantum entanglement below: the notion of the universe evolving on some uncomputable fractal invariant set I_U.

6.3.2   Chaos Theory and Quantum Theory

One seeming obstacle between quantum theory and chaos theory—the linearity of the former and the nonlinearity of the latter—is easily overcome. Figure 6.3 shows, using Monte Carlo techniques based on the Lorenz equations, the evolution of some contour of a probability distribution. The evolution of probability density ρ in classical physics satisfies a linear Liouville equation which in Hamiltonian form can be written

$$\frac{\partial \rho}{\partial t} = \{H, \rho\} \tag{6.2}$$

where $\{\ldots\}$ is the Poisson bracket. This is remarkably close in structure to the von Neumann-Dirac form

$$i\hbar \frac{\partial \rho}{\partial t} = [\mathbf{H}, \rho] \tag{6.3}$$

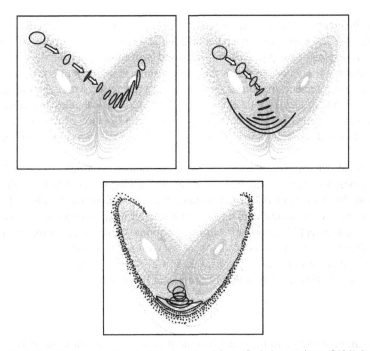

Fig. 6.3 Evolution of a contour of probability, based on Monte Carlo integrations of (6.1), is shown evolving in state space for different initial conditions, with the Lorenz attractor as background. The linearity of the Liouville equation exists peaceably with the nonlinearity of the underlying dynamical equations

of the Schrödinger equation where [...] is the operator commutator. (A reason for the appearance of $i\hbar$ in (6.3) and not in (6.2) is discussed in the Appendix.) Now the linearity of the Liouville equation is simply a consequence of conservation of probability. In particular, the linearity of the Liouville equation says nothing whatsoever about the underlying nonlinearity of the dynamics which generates ρ. The close formal similarity between the Hamiltonian form of the Liouville equation and the von Neumann-Dirac equation is strongly suggestive (to the author at least) that there must also be some deterministic framework underpinning quantum physics. If this is so, then, by analogy with the Liouville equation, the linearity of the von Neumann-Dirac equation says nothing about the nonlinearity of this underpinning deterministic dynamic.

The principal obstacle in drawing together chaos and quantum theory is therefore not the linearity of the Schrödinger equation, but the Bell Theorem. Unless it explicitly violates local causality (i.e. is nonlocal), a conventional computable deterministic model of quantum spin must satisfy the Bell Inequality

$$C(0, 0) + C(0, 1) + C(1, 0) - C(1, 1) \leq 2 \qquad (6.4)$$

and thus be inconsistent with experiment. As usual, we imagine a source of entangled spin-1/2 particles prepared so that the total angular momentum of any pair of particles is zero. The spins of the particles are measured by remote experimenters Alice and Bob who can each choose to orient their measuring apparatuses in one of two ways (relative to a reference direction): conventionally these are referred to as $X = 0$, $X = 1$ (for Alice) and $Y = 0$, $Y = 1$ (for Bob). $C(X, Y)$ denotes the correlation in spin measurements for ensembles of particles, as a function of the measurement settings. In the discussion below, $X' = 1$ if $X = 0$ and vice versa, and similarly for Y.

In a local deterministic theory, each pair of entangled particles is described by a supplementary variable λ, often referred to as a hidden variable (though in the uncomputable type of theory proposed here, there is no need for λ to be hidden; as discussed below, the ontic characteristics of a putative quantum state are inaccessible to the experimenter [22]). For each λ in a conventional hidden-variable theory, a value of spin (here ± 1) is defined for each of the four values of X and Y.

One property which a locally causal deterministic theory must conform to if it is to satisfy the Bell Inequality is that of Statistical Independence

$$\rho(\lambda | X Y) = \rho(\lambda) \tag{6.5}$$

The assumption (6.5) ensures that when the individual correlations in (6.4) are estimated from separate sub-ensembles of particle pairs (as happens in any real-world experimental test of the Bell inequality), then the hidden variables associated with these sub-ensembles are statistically equivalent to one another. A theory which violates (6.5) is referred to as superdetermistic [6], and the existence of statistically inequivalent real-world sub-ensembles is almost universally seen as implausibly conspiratorial and even unscientific.

However, in a non-computable theory, it is possible to violate (6.5) without negating the statistical equivalence of real-world sub-ensembles of particles/hidden variables. To see this, suppose indeed that the universe is evolving on some uncomputable fractal invariant set I_U in cosmological state space, as discussed above. Hence, if a pair of entangled particles, represented by some unique λ, is measured with some particular choice of measurement settings (X, Y), then, by hypothesis, the state $\mathbf{x}(\lambda, X, Y)$ of the world associated with the triple (λ, X, Y) lies on I_U.

Now if $\mathbf{x}(\lambda, X, Y)$ is a real-world state, then $\mathbf{x}(\lambda, X', Y)$ or $\mathbf{x}(\lambda, X, Y')$ are counterfactual states: they describe hypothetical worlds where measurement pairs such as $(X'Y)$ or (XY') might putatively have been performed on the same particle pair, even though the measurements (X, Y) were performed in reality. An experimenter might whimsically have decided to use the parity of yesterday's Dow-Jones index to determine whether to choose X over X'. In this way the counterfactual $\mathbf{x}(\lambda, X', Y)$ lies on I_U if it is consistent with the laws of physics to perturb the degrees of freedom that determined yesterday's Dow Jones index, and keep fixed all other degrees of freedom—the position of the moons of Jupiter and so on—that describe the state of the world. Now it is important to note that such a perturbation as formulated is merely kinematic perturbation and may or may not be consistent with constraints associated with the dynamical laws. Now as discussed above, a random dynamically unconstrained perturbation to a point on a Cantor set almost certainly takes the point off the Cantor Set. Hence in a model where states are constrained to lie on a measure-zero fractal invariant set in state space, it is plausible that neither the counterfactual states $\mathbf{x}(\lambda, X', Y)$ nor $\mathbf{x}(\lambda, X, Y')$ lie on I_U. In the Appendix we describe a particular model where such counterfactual states *definitely* do not lie on I_U. That is to say

$$\rho(\lambda|XY) = 1 \implies \rho(\lambda|X'Y) = 0 \text{ and } \rho(\lambda|XY') = 0 \qquad (6.6)$$

which is a manifest violation of (6.5).

On the other hand, since it is undecidable whether $\mathbf{x}(\lambda, X, Y) \in I_U$, then there is no algorithm for determining which of $\mathbf{x}(\lambda, X, Y)$, $\mathbf{x}(\lambda, X', Y)$, $\mathbf{x}(\lambda, X, Y')$ or $\mathbf{x}(\lambda, X', Y')$ lies on I_U. That is to say, we cannot assert by algorithm that $\rho(\lambda|XY) = 1$ and $\rho(\lambda|X'Y) = 0$. From a computational perspective, it is no more likely that the particle pair associated with a particular λ is measured with one set of X and Y values as with any other set. From this computational perspective, (6.5) is not violated.

Violations of Statistical Independence are sometimes seen as implausibly con-spiratorial. Indeed, it is sometimes said that it would be impossible to do science in a world where Statistical Independence is violated. For example, how could one test the efficacy of a new drug, if one could not guarantee that the sub-samples of volunteers given the drug and a placebo were not statistically equivalent? However, such criticisms simply do not apply to this violation of Statistical Independence, simply because by construction, the violations do not apply to anything happening in the real world, but only to counterfactual worlds. If both sub-samples of volunteers are sub-samples in the real world, then their statistical properties are by definition totally unaffected by the violation of Statistical Independence as described by (6.6). Similarly, such violation does not imply any conspiracy in the real world, because nothing the events that are being denied as unphysical, by construction do not occur in the real world.

Of course this result does not imply that all counterfactual worlds are physically inconsistent. Indeed there can be as many allowed counterfactual worlds on the invariant set as there are points on the real line. Hence, none of the analysis above denies the validity of counterfactual reasoning in classical physics, or indeed in everyday life. It is just that counterfactual reasoning fails when it comes to arguing about certain types of experiment in quantum physics.

In this analysis, it is important to note that we are invoking the concept of undecid-ability in relation to the geometry of state space, and not in terms of physical processes occurring in space-time. A mathematical description of an actual process occurring in physical space-time should always be possible using computational/algorithmic equations, whilst a mathematical description of a proposition concerning the reality of a putative state in state space may not be so. To illustrate this dichotomy, let us return to the prototype Lorenz attractor $\mathcal{A}_L$. As discussed, the proposition $\mathbf{x} \in \mathcal{A}_L$ is undecidable. However, the space-time processes which $\mathcal{A}_L$ represents, in this case a highly truncated model of fluid convection in Newtonian space-time, are given by the computable differential equations (6.1), solvable by algorithm.

Unravelling the dichotomy between real-world processes in space-time and putative worlds in state space is central to understanding why an uncomputable theory of quantum physics can violate Bell inequalities without violating experimenter free will or causality [6]. Free will is frequently described as an ability 'to have done otherwise', a description that is manifestly built around the notion of putative counterfactual worlds in state space (where I did do otherwise). However, one can equally well describe free will solely in terms of real-world processes, without referring to counterfactuals at all: specifically one is free when there are no constraints preventing one from doing as one wishes [8]. Similarly for causality: when Newton claps his hands and hears the sound reflected from the back wall of the college quad, he can assert that the sound was caused by the clap in one of two ways: either by claiming that if he hadn't clapped he wouldn't have heard the sound (thus invoking counterfactuals) or by linking the clap to the excitation of an acoustic wave which propagated across the quad according to the computational equations of fluid mechanics, was reflected by the quad wall, entered his ear triggering an electric signal in his neurons. The latter description only involves real-world processes occurring in space-time.

In the case of an uncomputable theory of quantum physics, such as presented here, these two descriptions of free will and causality are profoundly inequivalent. By basing descriptions of free will and local causality strictly on computational processes occurring in space-time, and not on undecidable counterfactuals in state space, an uncomputable deterministic model need violate neither free will nor causality and still violate the Bell inequality. No conspiracies are needed to achieve this. The uncomputable model described briefly in the Appendix does this and more: using number-theoretic properties of trigonometric functions [14], it shows that whatever the choices that Alice and Bob actually make for X and Y, the correlations for the sub-ensembles are *necessarily* quantum mechanical in nature.

Above, it was mentioned that in state-space, the p-adic metric respects the primacy of a fractal invariant set better than does the familiar Euclidean metric. This has profound metaphysical implications. For example, in the theory of counterfactual causality of the renowned philosopher David Lewis [10], it is assumed that of two putative counterfactual worlds, the one which resembles reality better must be closer to reality. From this perspective, a world which differs only in some seemingly insignificant detail, e.g. in the wavelength of a single photon from a distant quasar, must be extremely close to reality. As such, it would seem grossly implausible of a theory to assert as profoundly unphysical, a world which differs from reality only in terms of something as seemingly insignificant as the wavelength of a single photon. However, relative to the p-adic metric, if this counterfactual world lies in a fractal gap of the corresponding invariant set in state space, then no matter how closely it resembles reality, and hence no matter how close it appears to reality from a Euclidean perspective, it will actually be distant from reality. In the Appendix is discussed a model where such a counterfactual world does indeed lie in a fractal gap, if the wavelength of the photon is used to determine the measurement settings in a Bell experiment.

The results above carry over *mutatis mutandis* to the finite case where propositions about states lying on periodic fractal-like invariant sets may be formally decidable but nevertheless computationally irreducible (a term defined above). In such a case, it is impossible to determine reliably the truth of such propositions $\mathbf{x} \in I_U$ using a computational subset of the universe (e.g. what we would call a computer!). The model described in the Appendix has a finite but computationally irreducible representation for finite fractal parameter p. In this model, the continuum complex Hilbert space of quantum theory arises as a singular (and not a smooth) limit [1] at $p = \infty$. This is consistent with emerging evidence from quantum complexity theory [7] that quantum theory cannot be considered the smooth limit of some corresponding finite theory.

What is the key difference between a classical deterministic system described probabilistically by (6.2) and a quantum system described by (6.3)? In the classical system, the individual state-space trajectories are independent of one another: each trajectory has been started from an independently chosen set of initial conditions. By contrast, if the laws of physics derive from a constraint that trajectories must lie on some fractal geometry in state space then state-space trajectories can no longer be considered independent. Indeed, the primacy of such geometry implies that neighbouring state-space trajectories can be thought of as interacting with one another. A manifestation of such interaction is the quantum computer: a physical object able to perform certain computations exponentially faster than a classical computer. Such an ability is consistent with the notion that the evolution of physical objects associated with "reality" can be influenced by counterfactual processes occurring on neighbouring state-space trajectories on the invariant set.

6.3.3 Quantum Theory and General Relativity Theory

Could non-computability break the road-block in finding a satisfactory theory which can synthesise quantum and gravitational physics? Geroch and Hartle [5] and Penrose [20] have speculated that a quantum theory of gravity may not be computable on the basis that it is undecidable whether two simplicial 4 manifolds are topologically equivalent. The following additional reasons suggest that non-computability could lie at the heart of a quantum theory of gravity. Specifically:

- General relativity is a nonlinear theory. The structure of fractal geometry is necessarily nonlinear. As discussed, it is only when expressed in terms of evolution of probability that fractal-based dynamics, like the Schrödinger equation, appears linear.
- General relativity is a deterministic causal theory. Causal structure derives from the metric properties of space-time. As described in this essay, it is possible to violate Bell inequalities with a locally causal but uncomputable deterministic theory, providing the notion of local causality is defined purely in terms of computational processes occurring in space-time, and not in terms of undecidable counterfactual properties of state space.

- General relativity is primarily a geometric theory. Non-computability has a natural expression in terms of fractal (p-adic) geometry. Could it be that the computational pseudo-Riemannian geometry of space-time is emergent from the non-computational p-adic geometry of state space for large but finite p? (In particular, could it be that the Lorentzian signature of the space-time metric is emergent from the primitive quaternionic structure that is a feature of the specific fractal model discussed in the Appendix [18].)

6.4 Discussion

From where do new ideas come? Do they pop out of the aether as some random flashes of inspiration with no obvious precedent? Or do these ideas mostly already exist, but in a completely separate setting. As such, does the creative spark really consists of taking some pre-existing idea from its usual setting and transplanting it into an unfamiliar setting where it may provide new insights into old unsolved problems? Here, an example of the latter is presented. For many decades of his research career, the author has worked on the chaotic dynamics of climate, having first done a PhD in

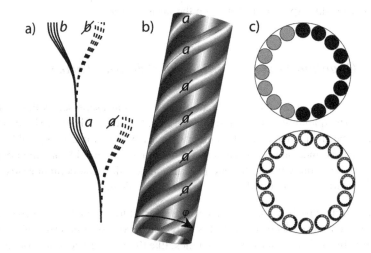

Fig. 6.4 A schematic of the local fractal state-space structure of the invariant set I_U in Invariant Set Theory. **a** An ensemble of trajectories decoheres into two distinct clusters labelled a and $\not{a}$. Under a second phase of decoherence, this trajectory, itself comprising a further ensemble, decoheres into two further distinct regions labelled b and $\not{b}$. **b** Under magnification, a trajectory segment is found to comprise a helix of p trajectories at the next fractal iterate. **c** Top: a cross section through the helix of trajectories comprises p (here $p = 16$) disks coloured black or grey according to whether that trajectory evolves to the a cluster or the $\not{a}$ cluster. Bottom: each of these p disks itself comprises p further disks coloured black or grey according to whether each trajectory evolves to the b or $\not{b}$. The fractal set C_p of disks is homeomorphic to the set of p-adic integers

general relativity theory. It was this somewhat improbable combination of research topics that led to a realisation [16] that the non-computable state-space geometry of chaotic systems could provide new insights into the Bell Theorem. Perhaps this could, in turn, provide a novel path to a unification of the two most fundamental theories of 20th century physics: general relativity theory and quantum theory. If this path does turn out to be the right one, quantum theory will have to change much more than general relativity.

1 Appendix

In Invariant Set Theory (IST) [17, 18], we consider a specific fractal model I_U of state-space trajectories (or histories), where a single trajectory at some $I - 1$th level of fractal iterate comprises a helix of p trajectories at the Ith fractal iterate (similar to a strand of rope)—see Fig. 6.4. Here p is a finite but arbitrarily large integer. A cross-section of such trajectories is isomorphic to C_p (see Fig. 6.2b). Under interaction with the environment, these Ith trajectories diverge (with the appearance of Everettian branching at the $I - 1$th iterate). In the simplest case where we partition state space into two clustering regions (corresponding to measurement eigenstates) labelled a and ϕ, each of the Ith-iterate trajectories is labelled by the region into which the trajectory evolves. In this way, at each fractal iterate, the set of p trajectories can be represented by the complex Hilbert vector $\cos\frac{\theta}{2}|a\rangle + e^{i\phi}\sin\frac{\theta}{2}|\phi\rangle$ where $\cos^2\frac{\theta}{2}$ denotes the fraction of Ith-iterate trajectories labelled a, and ϕ denotes an angular coordinate around the helix of Ith-iterate trajectories. This correspondence with complex Hilbert states only holds when $\cos\theta$ is of the form $\frac{n_1}{p} \in \mathbb{Q}$ and $\frac{\phi}{2\pi}$ is of the form $\frac{n_2}{p} \in \mathbb{Q}$ (where $0 \le n_1, n_2 \le p$ are integers). The finite size of the helix, and its symmetry under discrete rotations, explains both the factor $\hbar$ (whose dimensions are that of phase space) and i in (6.3). Indeed, as discussed in [18], I_U exhibits a natural quaternionic structure (generating Pauli spin matrices). In IST the state space S_p associated with such qubit Hilbert vectors is a discretised form of the Bloch Sphere.

Complementarity in quantum theory (arising from the non-commutativity of observables) is a consequence of number theory in IST. In particular Niven's Theorem [14] asserts that if ϕ is a rational angle, then $\cos\phi$ is almost certainly irrational. Such a transformation arises in performing a unitary Hadamard on a complex Hilbert vector. In IST, this number theoretic result implies that if an experimenter measures on which arm a particle travels when passing through an interferometer (position measurement), the experimenter could not counterfactually have performed an interferometric experiment (momentum measurement) on that same particle. Because of Niven's Theorem, S_p does not map onto itself under a general rotation of the sphere. That is, S_p is incommensurate under general rotations. In IST, the tensor product of rational Hilbert vectors is simply the Cartesian product of S_p—the discretised form implying that an exponentially increasing number of degrees of freedom can be accommodated with multiple Cartesian products, unlike with the continuum state space $\mathbb{S}^2$ of a qubit in quantum theory. As a consequence, for a Bell State, IST demands that the cosine of the relative angle between Alice and Bob's measurement settings must be of the form $\frac{n_3}{p}$ and hence be rational.

Consider the first two terms in the CHSH inequality (6.4). They refer to correlations relative to the three measurement orientations $X = 0$, $Y = 0$ and $Y = 1$. These can be represented as vertices of a spherical triangle on the celestial sphere. Suppose in reality that a particular particle pair (with hidden variable λ) were measured relative to $X = 0$, $Y = 0$. From IST, the cosine of the angular distance between the vertices $X = 0$ and $Y = 0$ must be rational. In [18], it is shown, using the cosine rule for spherical triangles, that in such a circumstance it is impossible for the cosine of the angular distance between the vertices $X = 1$ and $Y = 1$ to also be rational. Hence if the state $\mathbf{x}(\lambda, X = 0, Y = 0) \in I_U$, the counterfactual state $\mathbf{x}(\lambda, X = 0, Y = 1) \notin I_U$ (consistent with (6.6)). Because the fractal geometry is described using complex Hilbert vectors, the correlations between sub-ensembles of real-world particle pairs will necessarily be quantum mechanical.

A corresponding analysis can be made of the Factorisation assumption in the Bell Theorem, expressed as $A_{XY}(\lambda) = A_X(\lambda)$, $B_{YX}(\lambda) = B_Y(\lambda)$ where A and B are Alice and Bob's deterministic spin functions which return the values ± 1. Violation of Factorisation is typically viewed as implying a violation of local causality. However, in a geometric uncomputable theory, it is again vital to make the distinction between violation of local causality based on computable space-time processes—for example when information in space-time propagates superluminally—and violation of local causality based on undecidable counterfactual reasoning. As with free choice, it is possible to violate the latter without violating the former.

IST violates Factorisation from the counterfactual perspective, but not from the space-time perspective. From the space-time perspective if λ is associated with a state on I_U, then either $\lambda \in \Lambda_1 = \{\lambda | X = 0, Y = 0 \text{ or } X = 1, Y = 1\}$ or $\lambda \in \Lambda_2 = \{\lambda | X = 0, Y = 1 \text{ or } X = 1, Y = 0\}$. If $\lambda \in \Lambda_1$, then given X (say $X = 0$) the value of Y is redundant (it is necessarily $Y = 0$), and hence Factorisation is satisfied. Similarly if $\lambda \in \Lambda_2$. By contrast, Factorisation is violated if we admit counterfactual worlds not lying on I_U. Such a counterfactual world would be associated with one where, for example, $\lambda \in \Lambda_1$ and, with $X = 0$ fixed, $Y = 0$ is mathematically perturbed to $Y = 1$. Such a counterfactual state does not lie on I_U and is therefore, by the Invariant Set Postulate, not ontic. Such a state is, moreover, p-adically distant from any state on I_U (even if the counterfactual world only differs by the wavelength of a single photon, e.g. from a distant quasar). As such, IST is locally causal from a space-time perspective, but not from a counterfactual perspective.

The uncomputable theory described here is not quantum theory. However, the closed complex Hilbert Space of quantum theory is emergent (only) in the singular limit at $p = \infty$. At this limit, all fractal gaps close, and the set of all complex Hilbert states is ontic. As the theoretical physicist Michael Berry [1] has discussed, old theories of physics are frequently the singular limit of new theories as some parameter of the new theory (here p) is set equal to either zero or infinity.

References

1. M. Berry, Singular limits. Phys. Today **55**, 10–11 (2002)
2. L. Blum, F. Cucker, M. Shub, S. Smale, *Complexity and Real Computation* (Springer, 1997)
3. N.J. Cornish, Fractals and symbolic dynamics as invariant descriptors of chaos in general relativity, arXiv.org/gr-qc/9709036 (1997)
4. S. Dube, Undecidable problems in fractal geometry. Complex Syst. **7**, 423–444 (1993)
5. R. Geroch, J.B. Hartle, Computability and physical theories. Found. Phys. **16**, 533 (1990)
6. S. Hossenfelder, T.N. Palmer, Rethinking superdeterminism. Front. Phys. (to appear), arXiv:1912.06462 (2019)
7. M. Ji, A. Natarajan, T. Vidick, J. Wright, H. Yuen, MIP*=RE, arXiv:2001.04383 (2020)
8. R. Kane, *Free Will* (Blackwell, 2002)
9. S. Katok, *P-adic Analysis Compared with Real* (American Mathematical Society, 2007)
10. D. Lewis, Causation. J. Philos. **70**, 556–567 (1973)
11. E.N. Lorenz, Deterministic nonperiodic flow. J. Atmos. Sci. **20**, 130–141 (1963)
12. E.N. Lorenz, The predictability of a flow which possesses many scales of motion. Tellus **21**, 289–307 (1969)
13. A.E. Motter, Relativistic chaos is coordinate invariant. Phys. Rev. Lett. **91**, 231101 (2003)
14. I. Niven, *Irrational Numbers* (The Mathematical Association of America, 1956)
15. E. Ott, *Chaos in Dynamical Systems* (Cambridge University Press, 2002)
16. T.N. Palmer, A local deterministic model of quantum spin measurement. Proc. Roy. Soc. **A451**, 585–608 (1995)
17. T.N. Palmer, The invariant set postulate: a new geometric framework for the foundations of quantum theory and the role played by gravity. Proc. Roy. Soc. **A465**, 3165–3185 (2009)
18. T.N. Palmer, Discretization of the Bloch sphere, fractal invariant sets and Bell's theorem. Proc. Roy. Soc. https://doi.org/10.1098/rspa.2019.0350, arXiv:1804.01734 (2020)
19. T.N. Palmer, A. Doering, G. Seregin, The real butterfly effect. Nonlinearity **27**, 123–141 (2014)
20. R. Penrose, *The Large, the Small and the Human Mind* (Cambridge University Press, 1997)
21. A.M. Turing, On computable numbers, with an application to the entscheidungsproblem. Proc. Lond. Math. Soc. **42**, 230–265 (1981)
22. J. Walleczek, Agent inaccessibility as a fundamental principle in quantum mechanics: objective unpredictability and formal uncomputability. Entropy **21**, 4 (2019)
23. C.F. Woodcock, N.P. Smart, p-adic chaos and random number generation. Exp. Math. **7**, 333–342 (1998)

Chapter 7
A Gödelian Hunch from Quantum Theory

Hippolyte Dourdent

7.1 Introduction

In classical logic, self-referring propositions can lead to pathologies such as the
well-known Liar paradox "This sentence is false." Because it features an over-
determination—if the sentence is true then it is false, if it is false then it is true—the
"Liar" leads to undecidability, the impossibility to decide whether the sentence is
true or false. Analogs have been famously used in the foundations of mathematical
logic, from Russell's paradox to Gödel's incompleteness theorem, passing by Tarski
and Gödel undefinability theorem.[1]

In [1], Szangolies points out that "an intriguing connection between fundamental features of quantum mechanics and the phenomena of self-reference" might be
established. The expression "Gödelian hunch" is coined to describe "the idea that
the origin of the peculiarities surrounding quantum theory lie in phenomena related,
or at least similar, to that of incompleteness in formal systems." *What if the paradoxical nature of quantum theory could find its source in some undecidability analog
to the one emerging from the Liar ?* This essay aims at arguing for such quantum
Gödelian hunch via two case studies: quantum contextuality as an instance of the
Liar-like logical structure of quantum propositions; and the measurement problem
as a self-referential problem.

Quantum contextuality results from a theorem established by Kochen and Specker
[2], which shows that a quantum measurement cannot reveal a pre-existing value of
a measured property independently of the measurement context. Using a narrative
based on the Newcomb problem [3], the theological motivational origin of this result

---

[1]The undefinability theorem stipulates that any description of the truth of a proposition must be in
a richer metalanguage than the language in which the proposition itself is stated; this hierarchy of
languages arising as a solution of the Liar.

H. Dourdent (✉)
CNRS, Grenoble INP, Institut Néel, Université Grenoble Alpes, 38000 Grenoble, France
e-mail: hippolyte.lazourenko-dourdent@neel.cnrs.fr

© The Author(s), under exclusive license to Springer Nature Switzerland AG 2021   97
A. Aguirre et al. (eds.), *Undecidability, Uncomputability, and Unpredictability*,
The Frontiers Collection, https://doi.org/10.1007/978-3-030-70354-7_7

is introduced in order to show how the theorem might be related to a Liar-like undecidability (Sect. 7.2). I will also briefly present a topological generalization of contextuality [4] such that non-locality (Bell's theorem [5]) can be treated as a special case. In this approach, the logical structure of quantum contextuality is compared to sequences of cyclically referring statements, *"Liar cycles"*, which, associated with a truth predicate, lead to a logical contradiction [6].

The measurement problem is often presented as a tension between the linear and deterministic evolution of the wave-function following the Schrödinger equation and the projection postulate or the Born rule. Nevertheless, the problem was also analyzed as emerging from a logical error, and occurs because no distinction is made between theoretical and meta-theoretical objects. I will present my analysis of the Wigner's friend thought experiment [7] and a recent paradox by Frauchiger and Renner [8], introducing the notion of "meta-contextuality" as a Liar-like feature underlying the neo-Copenhagen interpretations of quantum theory (Sect. 7.3).

Finally, this quantum Gödelian hunch opens a discussion of the paradoxical nature of quantum physics (Sect. 7.4) and the emergence of time itslef from self-contradiction (Sect. 7.5).

7.2 A Gödelian Hunch from Quantum Contextuality

In 1960, Specker submitted a paper entitled "Die Logik nicht gleichzeitig entscheid-barer Aussagen" [9] (translated as "The logic of propositions not simultaneously decidable" [10]). Inspired by Birkhoff and von Neumann's axiomatic approach to derive quantum theory from non-classical "experimental propositions" adapted to the experimental result of quantum mechanics, Specker asks: *"Is it possible to extend the description of a quantum mechanical system through the introduction of supplementary—fictitious—propositions in such a way that in the extended domain the classical propositional logic holds [...] ?"* The answer is negative, "except in the case of Hilbert spaces of dimension 1 and 2." A fruitful collaboration with Kochen will culminate in an enriched reformulation of Specker's result, today known as the Kochen-Specker theorem [2]. Thus, either a measurement reveals a pre-existing value of a measured property depending on the measurement context (quantum con-textuality), or such value is unpredictable[2] [12].

7.2.1 *Counterfactual Undecidability*

In his seminal work, Specker noticed an analogy between these simultaneously unde-cidable propositions of quantum theory and the undecidability of *counterfactual*

---

[2] For example the outcome might be brought-into-being by the act of measurement itself, "Unper-formed measurements have no results." [11]

propositions.[3] Hence, the question of an extension of quantum propositions in clas-sical logic is paralleled with:

"the scholastic speculations about the "Infuturabilien" [...], that is, the question whether the omniscience of God also extends to events that would have occurred in case something would have happened that did not happen. (cf. e.g. [3], Vol. 3, p. 363.)" [10]. Can an omniscience extend to counterfactual propositions ? A possible positive answer is given by the reference "([3], Vol. 3, p. 363) [13]". The latter leads to a chapter on *molinism*, describing an unorthodox form of omniscience proposed by scholastics in order to conciliate God's foreknowledge and human's free will. According to this view, God's knowledge of counterfactual facts, i.e. facts condi-tioned on our free choices, precedes God's knowledge of actual facts. God already has knowledge of our free acts, but our free acts have a counterfactual power on his knowledge. If God had predicted that you will make a certain choice A, it may nevertheless have been in your power to do something, such that were you to do it, God would not have predicted this peculiar choice A. In a sense, God's omniscience and human free will can co-exist at the condition that the former is *contextualized* by the latter.

In order to illustrate the afored mentioned analogy, I propose the following narra-tive. We invoke an omniscient demon whose omniscience extends to counterfactual propositions. Two observables A and B are given to a free agent, Alice. Alice can choose to measure the observable B in two contexts: $C_1 := (A, B)$ or $C_2 := (B)$. Beforehand, the demon has predicted her choice and, based on it, has assigned a value to B: $v(B)_{|C_1} = 0$ or $v(B)_{|C_2} = 1$. Alice measures the value of the observable in one of the contexts, and assume that she verifies that the demon's prediction is correct. One can then ask the counterfactual question: what would have happened if she had chosen the other context ? Two solutions are possible:

- (a) "If Alice had chosen the other context, she would have found a different value for B." In this case, the omniscience of the demon may extend to counterfactuals. But this implies that either Alice is not free of her choices (superdeterminism), or the omniscience of the demon is conditioned by the context she chooses (molin-ism).
- (b) "If Alice had chosen the other context, she would have found the same value for B." In this case, the omniscience of the demon does not extend to counterfactuals. The demon would have been wrong. Because its essence is defined by its function, denying this function is an exorcism. Thus, the value of B is *unpredictable*.

This narrative is freely inspired by the Newcomb problem [3], a decision theory problem where the values associated to the prediction correspond to distinct earnings (e.g. $v(A) = 10k\$$, $v(B)_{|C_1} = 1k\$$ and $v(B)_{|C_2} = 1M\$$), the problem arising from the question of which choice allows Alice to maximize her gains. Interestingly, Slezak observed that the problem might originate from a self-referential structure: *"Newcomb's problem may be understood as a game against one's self in which one's*

---

[3] A counterfactual proposition is a special kind of conditional proposition which follows the struc-ture: "If A' would have happened instead of A, then B' would have happened instead of B."

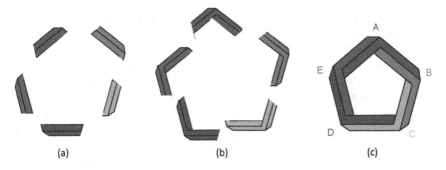

Fig. 7.1 a Each corner of an individual bar represents an observable to which one assign a truth-value. **b** Each observable is compatible with two other ones separately, and thus two local contexts can be defined per observable. The truth values assigned to observables in a context are logically consistent. **c** Each corner from $\{A, B, C, D, E\}$ is mutually compatible with its two neighbours. However, the global picture of all bars glued together is an undecidable figure, the Penrose pentagone. One cannot define a global context in which no truth-value assignment leads to a contradiction

choice is based on deliberations that attempt to incorporate the outcome of this very choice. [...] This hidden circularity facing the decision-maker arises because, as we contemplate our best move, we consider the demon's decision, which is actually based on this very choice we are trying to make." [14]

A similar "circularity" lies under the counterfactual statements (a) and (b). It is of course trivial to point out that nothing is quantum in the Newcomb narrative. Yet, the non-Boolean logical structure of quantum theory yields analog conclusions: either a value-assignement to all observables is contextual or one cannot assign pre-defined values to all observables, i.e. these values are in general unpredictable. The self-referential nature of these narratives hints at the presence of a similar circular structure underlying quantum contextuality. Approaching contextuality as the fact that quantum theory is based on intertwined Boolean algebras that cannot be embedded in a global Boolean algebra highlights this Liar-like structure.

7.2.2 Topological Undecidability

In a topological approach of contextuality by Abramsky et al. [4] based on sheaf theory and cohomology, contextuality emerges when data which are locally consistent are globally inconsistent. One can illustrate this definition of contextuality with famous undecidable figures such as the Penrose pentagone[4] (Fig. 7.1). In this construction, each pair of bars can be isolated and visualized without paradoxes. It is only when one tries to interpret the figure globally that a visual obstruction emerges.

---

[4] Warning: this is a figurative illustration which has a didactic purpose. Of course, sheaf-theoretic contextuality cannot be reduce to this simple example. However, note that a proof of contextuality, the violation of the KCBS inequality [15], shares a similar structure.

As noticed in [6], there is a direct connection between contextuality and classical semantic paradoxes called "Liar cycles", defined as sequences of statements of the form : $[\{S_1, S_2\}$ true ; ... ; $\{S_{N-1}, S_N\}$ true ; $\{S_N, S_1\}$ false] with S_i the ith assertion, and $\{S_{i-1}, S_i\}$ and $\{S_i, S_{i+1}\}$ the two "local" contexts associated to this assertion. Although every proof of the Kochen-Specker theorem features such logical global obstruction, this generalized approach does not reduce to quantum contextuality, and also incorporates non-locality as a special case. As an example, the Hardy paradox [16] can be shown to entail contextuality, and thus feature a Liar-like logical structure. The paradox involves two agents, Alice and Bob, who share a two-qubit system in a specific entangled state. Each agent can choose to measure their respective qubit in a computational $\{|0\rangle, |1\rangle\}$ or a diagonal basis $\{|+\rangle, |-\rangle\}$ with $|\pm\rangle = \frac{1}{\sqrt{2}}(|0\rangle \pm |1\rangle)$. The initial entangled state can thus be written in four different basis, each corresponding to a measurement context. For example, in the comput.-comput. basis, the state is: $|\psi\rangle = \frac{1}{\sqrt{3}}(|00\rangle + |10\rangle + |11\rangle)$. Assuming that a predefined value can be associated to a measured property when a result can be predicted with certainty, one can infer the four following sentences, each associated to a measurement context (cf. detailed construction in appendix):

Sentence H1: "If Alice obtains '$-$', then Bob obtains '1'." (diago.-comput. basis)

Sentence H2: "If Bob obtains '1', then Alice obtains '1'." (comput.-comput. basis)

Sentence H3: "If Alice obtains '1', then Bob obtains '+'." (comput.-diago. basis)

Sentence H4: "Alice and Bob both obtain '$-$' with a probability 1/12.". (diago.-diago. basis)

Assuming non-contextuality means that one can build inferences from these different sentences. For instance, from $(H1, H2, H3)$, one can construct the sentence: "If Alice obtains '$-$', then Bob obtains '+' ". However, this sentence is incompatible with $H4$. Thus, $((H1, H2, H3), H4)$ is globally inconsistent, and the paradox entails contextuality. The following probabilistic[5] Liar cycle can be formulated, assuming that both Alice and Bob obtained '$-$': Bob obtains '$-$' and Alice obtains '$-$' $\rightarrow$ Bob obtains '1' $\rightarrow$ Alice obtains '1' $\rightarrow$ Bob obtains '+', contradicting the first assignment. Note that in such contextuality scenario, the contradiction occurs at the level of classical statements, inferred from quantum propositions. The assigned values are both classical and meta-theoretical, in the sense that they are not part of quantum theory. Hence, if one wants to attach meta-theoretical statements to quantum propositions, these statements cannot be embedded in a global Boolean one in general. The non-Boolean logic of quantum theory contaminates the meta-theoretical state-

---

[5] The Hardy paradox is a probabilistic Liar cycle because the contradiction only occurs with a probability 1/12.

ments, which become globally *undecidable*. I argue that this global undecidability of quantum propositions is in favor of a quantum Gödelian hunch.

7.3   A Gödelian Hunch from the Measurement Problem

As expressed in the literature, there exists different measurement problems (cf. e.g. [17]). The one we wish to tackle addresses "the question of what makes a measurement a measurement. [...] There is nothing in the theory to tell us which device in the laboratory corresponds to a unitary transformation and which to a projection !" [17]. This measurement problem as been analyzed as a "logical error" emerging from a a lack of distinction between theoretical and meta-theoretical objects [18]. Similar conclusions explicitly underlying an analogy between the measurement problem and Gödel's theorem have been made (cf. [1] for an overview). For example, Chiara notices that such analysis could seem "to be very close to some similar limitative results that we have accepted in logic such as the Gödel theorem (who realizes a proof of the consistency of a well-behaved scientific theory, must be 'external' with respect to the theory (in the sense that he cannot use only the proof theoretical tools allowed by the theory)) or the Tarski theorem (who 'grasps' the concept of truth for a well-behaved theory cannot speak only the language of the theory)." [19] I will analyze the Wigner's Friend thought experiment and the Frauchiger-Renner paradox -which shows that "a self-referential use of quantum theory yields contradictory claims." [8]—as sustaining this Gödelian hunch.

7.3.1   Wigner's Friend, Universality, Meta-Contextuality and Measurement

The measurement problem we are dealing with is usually formalized as follow. Assume that a quantum system is in the state $|\psi\rangle = \alpha|0\rangle + \beta|1\rangle \in \mathcal{H}_S$. On the one hand, following the projection postulate, the system will either be projected onto state $|0\rangle$ with probability $|\alpha|^2$, or state $|1\rangle$ with probability $|\beta|^2$ after the measurement. On the other hand, if the "observer" (e.g. the measuring device) is a physical system, then it shall be described by quantum theory. One associate a Hilbert space $\mathcal{H}_O$ to this observing system. Defining $|M\rangle$ the observer state "ready to perform a measurement", the initial state of the compound system in $\mathcal{H}_S \otimes \mathcal{H}_O$ is $(\alpha|0\rangle + \beta|1\rangle) \otimes |M\rangle$. In this case, the measurement process is described as an interaction between the system and the device, and thus as a unitary transformation U, resulting in $U[(\alpha|0\rangle + \beta|1\rangle) \otimes |M\rangle] \rightarrow \alpha|0\rangle \otimes |M_0\rangle + \beta|1\rangle \otimes |M_1\rangle$. Because the two final states are physically distinct, there seem to be a tension between the postulates of quantum theory, raising the question of how should a measurement process be described.

The Wigner's Friend thought experiment [7] is a meta-illustration of this measurement problem, which asks: what happens when an observer observes another observer observing a quantum system ? A quantum system, e.g. a qubit living in $\mathcal{H}_S$, is given to an observer, Wigner's friend, who can perform a measurement on this system in her laboratory. Outside her laboratory, another observer, Wigner, can associate a quantum state to the compound system $\mathcal{H}_S \otimes \mathcal{H}_O$, where $\mathcal{H}_O$ is a Hilbert space associated to Wigner's friend, e.g. a memory qubit $|M_i\rangle$ which can be interpreted as "Wigner's friend observes a projection on state $|i\rangle$". The problem occurs from the fact that while Wigner's friend observes a collapse of the qubit, the measurement process has been described as a unitary transformation from Wigner's perspective. However both descriptions should be valid. My analysis of this problem relies on the following terminology. The quantum system is an *object*, since it is described by quantum theory. Wigner's friend is an observer, and as a user of quantum theory, is a meta-theoretical object, in short a *meta-object*. Wigner is an observer who can perform a measurement on systems of the form *object* $\otimes$ *meta-object*, and is thus a meta-meta-object, or *meta-observer*. The problem seems to arise from the fact that an observer and a meta-observer are lead to describe the same event in contradictory ways. I introduce the notion of *meta-context* as a set of the form {meta-object,object}. This set is defined by a movable *cut* between theoretical objects studied in the language of the theory, and meta-theoretical objects which are out of the range of the theory. In the Wigner's friend paradox, two meta-context are involved: {Wigner's friend, $\mathcal{H}_S$} and {Wigner, $\mathcal{H}_S \otimes \mathcal{H}_O$}.

The problem can be understood as follows. Firstly, quantum theory is assumed to be correct and can be applied to any object whatsoever. Such assumption is called **quantum universality** (Q). Secondly, one assumes that truth values given by the propositions associated with an object are independent of the meta-context, of whether the object is theoretical or meta-theoretical, i.e. the truth values are **non-meta-contextual** (NMC).[6] Maintaining (Q) and (NMC) leads to an absolute form of universality: everything can be described by the theory, irrespective of the meta-context, no cut is needed. But imagine an infinite chain of observers observing observers observing a quantum system. Then, meta- ... -meta-observers are invoked, *ad infinitum*. One could argue that the ultimate meta$^\infty$-object is God, or some Laplacian demon. However, if such a demon can measure the whole Universe, then the demon is necessarily excluded from the Universe in order to avoid Liar-like inconsistencies, independently of the considered theory. As shown by Breuer [21], if a theory is considered to be absolutely universally valid, then the theory cannot be experimentally fully accessible, due to self-referential problems. There is a tension between absolute universality (Q,NMC), in which the measuring process might be treated theoretically, and **measurement** as a meta-theoretical process. In the light of this analysis, the most appealing solution is to drop (NMC) and acknowledge the observer for what it is: a meta-object. This way, the notion of meta-observer becomes obsolete, and the logical inconsistencies are avoided (cf. Fig. 7.2). The universality of the theory is maintained, but becomes *relative*. Any object can be cut

---

[6] This notion is equivalent to Brukner's "observer-independent facts" [20].

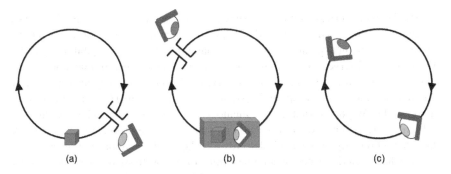

Fig. 7.2 Inspired by Grinbaum's epistemic loops [18], let us represent all theoretical objects by a loop. Cutting the loop sends objects in the meta-theoretical domain. **a** Meta-Context {Wigner's friend, $\mathcal{H}_S$}. **b** Meta Context {Wigner, $\mathcal{H}_S \otimes \mathcal{H}_O$}. **c** Maintaining (Q) and (MNC) leads to ignoring the relative cuts, i.e. the meta-contexts. Wigner and Wigner's friend are put at the same level, and self-referential inconsistencies may occur

and become a meta-object. However, once the cut is fixed, any out-of-meta-context question is **undecidable**. "Although it can describe *anything*, a quantum description cannot include *everything*." [11]

7.3.2 "Wigner's Friendifications"

Recently, there has been a renewed interest in Wigner's thought experiment in the field of quantum foundations. This resurgence is due to the appearance of new hybrid paradoxes [8, 20, 22], which rely on a Wigner's Friendification,[7] a transformation of previous quantum "paradoxes" where one allows meta-objects to be described as objects of the theory, and allows meta-observers to measure coumpound systems of the type "object ⊗ meta-object". I will analyze the Frauchiger-Renner paradox [8] as a Wigner's Friendification of the Hardy paradox explicitly showing the logical inconsistency which can emerge from (Q,MNC).

The original Hardy scenario involved two agents/observers, Alice and Bob, sharing a two-qubit system. In the new thought experiment, Alice and Bob are upgraded to meta-observers, while two new agents, their respective friends, share a two-qubit system and can perform a measurement on their respective part of the system. Like in the standard scenario, Alice and Bob's friend can measure their qubit in the computational {|0⟩, |1⟩} basis or in the diagonal {|+⟩, |−⟩} basis. Regarding Alice and Bob, these bases are "Wigner's friendified" as follows. The computational basis is transformed into a meta-computational basis corresponding to an "observer basis",

---

[7] In my knowledge, this terminology was first used by Aaronson in a blog post (www.scottaaronson.com/blog/?p=3975) in order to describe the Frauchiger-Renner paradox.

a statement made by the observer, the friend[8]: $\{|0\rangle_{S_A} \otimes |0\rangle_{F_A}, |1\rangle_{S_A} \otimes |1\rangle_{F_A}\}$. For example, if Alice's friend finds his qubit in state $|0\rangle_{S_A}$, then his statement will be $|0\rangle_{F_A}$ and Alice will find the global system in the state $|0\rangle_{S_A} \otimes |0\rangle_{F_A}$. The diagonal basis of the standard observation becomes a meta-diagonal basis corresponding to a "meta-observer basis", where the meta-observer actually performs a quantum measurement on the compound system, resulting in a statement associated to the meta-observer: $\{|+\rangle_A, |-\rangle_A\}$, with $|\pm\rangle_A = \frac{1}{\sqrt{2}}(|0\rangle_{S_A} \otimes |0\rangle_{F_A} \pm |1\rangle_{S_A} \otimes |1\rangle_{F_A})$. Applying this Wigner's Friendification to the four sentences of the Hardy paradox (cf appendix), one obtains four new assertions:

Sentence FR1: "If Alice obtains "$-$", then Bob's friend obtains outcome "1"."

Sentence FR2: "If Bob's friend obtains "1", then Alice's friend obtains outcome "1"."

Sentence FR3: "If Alice's friend obtains "1", then Bob obtains outcome "$+$"."

Sentence FR4: "Alice and Bob both obtain "$-$" with a probability of $\frac{1}{12}$."

Like in the Hardy paradox, these sentences forms a probabilistic Liar cycle: assume that Bob and Alice both obtains '$-$' (this happens with a probability 1/12). Bob obtains "$-$" and Alice obtains '$-$' $\rightarrow$ Bob's friend obtains "1" $\rightarrow$ Alice's friend obtains "1" $\rightarrow$ Bob obtains "$+$", contradicting the first statement. In [8], the authors analyze this paradox as an incompatibility between three assumptions: (Q) quantum theory is correct and can be applied to systems of any complexity; (C) observers and meta-observers claims should be consistent with each other; (S) a measurement yields a single outcome. Assumption (C), in particular, has been widely discussed in the literature (cf. for example [20, 23–25]). I argue that this assumption can be reformulated into two assumptions: non-contextuality and non-meta-contextuality.

Indeed, like the Hardy paradox, the Frauchiger-Renner paradox entails contextuality in the sense of Abramsky: a global logical obstruction of four consistent propositions.[9] Thus the contradiction might occur from assuming non-contextuality (NC). However, unlike the Hardy paradox, here each statement can be associated to one agent: one for each observer (FR2 and FR3), and one for each meta-observer (FR1 and FR4). In fact, like in the original Wigner's friend experiment meta-objects (the friends) are described in the language of the theory, i.e. at the level of objects. As seen previously, this is equivalent to the (NMC) assumption, which associated with (Q), can lead to self-referential inconsistencies when statements made in different meta-contexts are compared. Giving up on (NMC), consistency is restored, but only inside a meta-context among {Alice, Alice's Friend $\otimes$ qubit S_A} ; {Bob,

---

[8] More precisely, it corresponds to a meta-observer asking her friend in which state has the qubit been projected.

[9] Note that the paradox has already been analyzed as applying classical logic to quantum propositions which is forbidden by the non-Boolean structure of quantum theory [20, 23, 24].

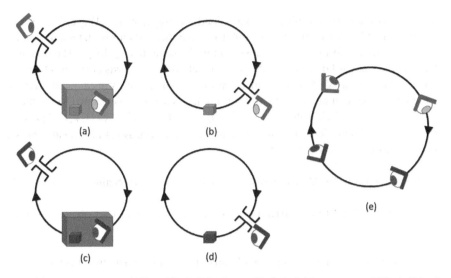

Fig. 7.3 **a** Meta-context: {Alice, Alice's Friend ⊗ qubit S_A}. **b** Meta-context: {Alice's Friend, qubit S_A}. **c** Meta-context: {Bob, Bob's Friend ⊗ qubit S_B}. **d** Meta-context: {Bob's Friend, qubit S_B}. **e** Maintaining (Q) and (NMC), i.e. comparing the results from different meta-contexts, leads to logical inconsistencies

Bob's Friend ⊗ qubit S_B} ; {Alice's Friend, qubit S_A} ; {Bob's Friend, qubit S_B} (cf. Fig. 7.3). Under such analysis, the fact that "a self-referential use of the theory yields contradictory claims" [8] is not especially surprising, if one acknowledge that quantum theory can only be consistently used in a meta-context, i.e. that *the use of quantum theory is (meta-)contextual*.

7.3.3 The Heirs of Copenhagen

Analyzing the measurement problem as self-reference and escaping the logical incon-sistency by introducing a *cut*[10] complies with various "neo-Copenhagen" interpre-tations of quantum theory, often wrongly labeled as "anti-realistic" [26], such as information-based interpretations [27–29] and QBism [30]. All agree on the fun-damental distinction between the meta-theoretical and theoretical object. In these interpretations, this movable cut is *functional* and not ontological. It does not dis-criminate a macroscopic classical world from a microscopic quantum one, because every object can be treated by the theory (Q) or not. This is especially made explicit in Rovelli's relational interpretation: "As soon as we realize that any physical sys-tem can play the role of a Copenhagen's 'observer', we fall into relational quantum mechanics. Relational quantum mechanics is Copenhagen quantum mechanics made

---

[10] Sometimes called the von Neumann or Heisenberg cut ("Schnitt").

democratic by bringing all systems onto the same footing." [29], as well as in a recent QBist analysis of the Frauchiger-Renner paradox, which rests on a: "quantum Copernican principle; when two agents take actions on each other, each agent has a dual role as a physical system for the other agent" [31]. Following the footsteps of Bohr: "There is no quantum world. There is only an abstract quantum physical description. It is wrong to think that the task of physics is to find out how nature *is*. Physics concerns what we can *say* about nature. We depend on our words, our task is to communicate experience and ideas to others. We are suspended in language ..." [32]; or as Wittgenstein wrote in his *Tractatus*: "(5.632) The subject does not belong to the world: rather it is a limit of the world." Absolute universality has a God-like flavour and leads to paradoxical features that cannot be *said*. On the contrary, one can acknowledge the transcendental status of the meta-theoretical object: a classical (Boolean) description is the condition of possibility for the rendering of quantum (non-Boolean) events.

7.4 Conclusion: Is Physics Paradoxical?

In his seminal paper on the logic of simultaneously undecidable propositions [9, 10], Specker attached the following epigraph: "La logique est d'abord une science naturelle." [Logic is in the first place a natural science.] extract from "La physique de l'objet quelconque" by Gonseth. Gonseth argued that logic should be considered as an experimentally refutable science of "any object whatsoever". If quantum physics goes against classical logic, thus classical logic should be revised. Several years later, Putnam defended a similar idea in a paper entitled "Is Logic Empirical ?" [33]. Mirroring this interrogation, we ask: "Is Physics Paradoxical ?".

Quantum theory does not only defy common sense, but it also defies classical logic, i.e. our common language and semantic. In this sense, quantum theory is more paradoxical than other physical theories. But is Nature itself paradoxical ? Does the world really feature intrinsically strange phenomena that cannot be grasped with our words, whether it is a non-local behaviour or parallel worlds ? In this essay, I argued for an alternative. Quantum paradoxes are not physical, but emerge from *a lack of metaphysical distancing*. I highlighted how the Liar-like structure of quantum propositions enlightened by the Kochen-Specker theorem already invites to give up on considering quantum objects as entities with intrinsic properties independently of the questions asked by a meta-theoretical object. I proposed the notion of "meta-contextuality" to explain how neo-Copenhagen interpretations avoid the measurement problem, Wigner's friend and Wigner's friendified paradoxes by analyzing them as logical errors. Acknowledging the need for an undiscriminating cut between meta-theoretical and theoretical objects when one uses quantum theory, any question that ignores this transcendental distinction looses its operational significance and becomes physically undecidable. Thus, quantum paradoxes might just be instances of a fundamental undecidability, contributing to a quantum Gödelian hunch.[11] Finally, this essay fully adheres to Wheeler's intuition[12]: "Physics is not machinery. Logic is not oil occasionally applied to that machinery. Instead, everything, physics included, derives from two parents, and is nothing but cathode-tube image of the interplay between them. One is the "participant". The other is the complex of undecidable propositions of mathematical logic." [26]

7.5 Epilogue: A Gödelian Hunch from Time

Quantum physics might not be the only branch of physics where one can hope to find physical analogs or instances of Liar-like paradoxes. In 1949, Gödel discovered solutions of general relativity latter known as closed time-like curves (CTCs) which theoretically would allow an observer to travel back in her own past [35]. However, the existence of such closed causal loops seems to imply the possibility for a traveller to interact with her own past-self, and for example prevent her own time-travel. This paradox, known as the grandfather antinomy, shares the same logical structure as the Liar. Unlike quantum theory, where the Gödelian hunch relies on the semantic of the theory, the grandfather paradox is a (speculative) *physical realization* of a self-contradiction.

By analogy with the scholastic debate previously introduced Sect. 7.2.1, the paradox can be understood as the tension between events that already happened and the ability to decide whether these "physically-already-determined" facts can be changed or not. Here, the role of God or the omniscient demon is played by *time* itself. The most popular (and boring) solution in science-fiction is a "many-worlds-like" one: there is no contradiction because when the traveler interacts with her past, different consistent worlds are created. One can defend a "superdeterminisitic" solution, where the traveler has no free will. A weaker version of this solution is that the traveler is still free, but her choices of actions are limited by some "time police / fine-tuning" principle (e.g. a Leibnizian notion of "compossible facts") such that consistency is preserved. Finally, one could deny time its fundamental aura, and argue instead that it is emergent. In fact, inside a closed loop, "time" is undefinable. Following the notion of contextuality introduced precedently, when one faces a global inconsistency, one can cut the loop, and recover logical consistency by introducing "local" contexts of logically consistent and well-defined sequences of events (cf. Fig. 7.4). These "contexts of ordered events" are locally consistent, but globally inconsistent.

---

[11] A very recent result [34] also contributes to the quantum Gödelian hunch. Using a modified proof of quantum contextuality, the authors proved that the class MIP* of problems that can be decided by a polynomial-time referee interacting with quantum agents sharing entanglement contains Liar-like undecidable problems.

[12] Wheeler might have been one of the first to investigate this quantum Gödelian hunch. A famous anecdote tells that Wheeler was thrown out of Gödel's office for asking him if there was a connection between his incompleteness theorem and Heisenberg's uncertainty principle [1].

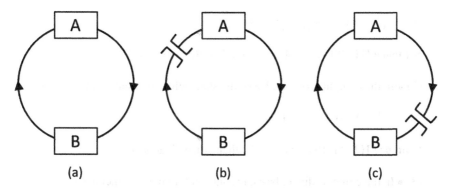

Fig. 7.4 a Events A and B in a closed loop. The order is undefinable. If the loop is cut, an order emerges. Depending on the position of the cut, the "context", either A precedes B (**b**), or the opposite (**c**)

As Gödel wrote: *"Time is the means by which God realized the inconceivable that P and non-P are both true [...]."* [36]

This way, time emerges from cutting self-referential paradoxes. Noticing that this cut might be epistemic, in line with a Gödelian hunch, one could finally speculate that *"Time is a consequence of every attempt to provide a comprehensive description of the universe from within. Thus, time in this sense is not related to the universe itself but to the attempt to describe it.'* [37]

Acknowledgements I would like to thanks Cyril Branciard for his precious support and advices, and Alexei Grinbaum for inspiring discussions.

Technical Appendices

The Hardy Paradox

In this scenario, two agents, Alice and Bob, share a two-qubits system in a specific entangled state. Each agent can choose to measure their respective qubit in a computational $\{|0\rangle, |1\rangle\}$ or a diagonal basis $\{|+\rangle, |-\rangle\}$ with $|\pm\rangle = \frac{1}{\sqrt{2}}(|0\rangle \pm |1\rangle)$. The initial entangled state can thus be written in four different basis, each corresponding to a measurement context. For example, in the comput.-comput. basis, the state is: $|\psi\rangle = \frac{1}{\sqrt{3}}(|00\rangle + |10\rangle + |11\rangle)$. Assuming that a predefined value can be associated to a measured property when a result can be predicted with certainty, one can infer the four following sentences, each associated to a measurement context:

(1) • In the diago.-comput. basis, the state before measurements is:

$$|\psi\rangle = \sqrt{\tfrac{2}{3}}|+0\rangle + \tfrac{1}{\sqrt{6}}|+1\rangle - \tfrac{1}{\sqrt{6}}|-1\rangle$$

Sentence H1 : "If Alice obtains '$-$', then Bob obtains '1'."

(2) • In the comput.-comput. basis, the state before measurements is:

$$|\psi\rangle = \tfrac{1}{\sqrt{3}}(|00\rangle + |10\rangle + |11\rangle)$$

Sentence H2 : "If Bob obtains '1', then Alice obtains '1'."

(3) • In the comput.-diago. basis, the state before measurements is:

$$|\psi\rangle = \sqrt{\tfrac{2}{3}}|1+\rangle + \tfrac{1}{\sqrt{6}}|0+\rangle + \tfrac{1}{\sqrt{6}}|0-\rangle$$

Sentence H3 : "If Alice obtains '1', then Bob obtains '+'."

(4) • In the diago.-diago. basis, the state before measurements is:

$$|\psi\rangle = \tfrac{3}{\sqrt{12}}|++\rangle + \tfrac{1}{\sqrt{12}}|+-\rangle - \tfrac{1}{\sqrt{12}}|-+\rangle + \tfrac{1}{\sqrt{12}}|--\rangle$$

Sentence H4 : "Alice and Bob can both obtain '$-$' with a probability 1/12."

Assuming non-contextuality means that one can build inferences from these different sentences. For instance, from $(H1, H2, H3)$, one can construct the sentence: "If Alice obtains '$-$', then Bob obtains '+' ". However, this sentence is incompatible with $H4$. Thus, $((H1, H2, H3), H4)$ is globally inconsistent, and the paradox entails contextuality. The following probabilistic[13] Liar cycle can be formulated, assuming that both Alice and Bob obtained '$-$': Bob obtains '$-$' and Alice obtains '$-$' $\rightarrow$ Bob obtains '1' $\rightarrow$ Alice obtains '1' $\rightarrow$ Bob obtains '+', contradicting the first assignment.

"Wigner's Friendification" of the Hardy Paradox

The Hardy paradox presented above is Wigner's friendified as follows: The computational basis is transformed into a meta-computational basis corresponding to an "observer basis" $\{|0\rangle_{S_A} \otimes |0\rangle_{F_A}, |1\rangle_{S_A} \otimes |1\rangle_{F_A}\}$. The diagonal basis of the standard observation becomes a meta-diagonal basis corresponding to a "meta-observer basis": $\{|+\rangle_A, |-\rangle_A\}$, with $|\pm\rangle_A = \tfrac{1}{\sqrt{2}}(|0\rangle_{S_A} \otimes |0\rangle_{F_A} \pm |1\rangle_{S_A} \otimes |1\rangle_{F_A})$. The corresponding sentences can then be derived:

---

[13] The Hardy paradox is a probabilistic Liar cycle because the contradiction only occurs with a probability 1/12.

(1) • In the metaobserver-observer basis, the state before measurements is:

$$|\psi\rangle_{tot} = \sqrt{\frac{2}{3}}|+\rangle_A|0\rangle_{S_B}|0\rangle_{F_B} + \frac{1}{\sqrt{6}}|+\rangle_A|1\rangle_{S_B}|1\rangle_{F_B} - \frac{1}{\sqrt{6}}|-\rangle_A|1\rangle_{S_B}|1\rangle_{F_B}$$

Sentence FR1: "If Alice finds the outcome '−', she knows that Bob's friend obtained outcome '1'."

(2) • In the observer-observer basis, the state before measurements is:

$$|\psi\rangle_{tot} = \frac{1}{\sqrt{3}}\left(|0\rangle_{S_A}|0\rangle_{F_A}|0\rangle_{S_B}|0\rangle_{F_B} + |1\rangle_{S_A}|1\rangle_{F_A}|0\rangle_{S_B}|0\rangle_{F_B} + |1\rangle_{S_A}|1\rangle_{F_A}|1\rangle_{S_B}|1\rangle_{F_B}\right)$$

Sentence FR2: "If Bob's friend finds the outcome '1', he knows that Alice's friend obtained outcome '1'."

(3) • In the observer-metaobserver basis, the state before measurements is:

$$|\psi\rangle_{tot} = \sqrt{\frac{2}{3}}|1\rangle_{S_A}|1\rangle_{F_A}|+\rangle_B + \frac{1}{\sqrt{6}}|0\rangle_{S_A}|0\rangle_{F_A}|+\rangle_B + \frac{1}{\sqrt{6}}|0\rangle_{S_A}|0\rangle_{F_A}|-\rangle_B$$

Sentence FR3: "If Alice's friend finds the outcome '1', she knows that Bob obtained outcome '+'."

(4) • In the metaobserver-metaobserver basis, the state before measurements is:

$$|\psi\rangle_{tot} = \frac{3}{\sqrt{12}}|+\rangle_A|+\rangle_B + \frac{1}{\sqrt{12}}|+\rangle_A|-\rangle_B - \frac{1}{\sqrt{12}}|-\rangle_A|+\rangle_B + \frac{1}{\sqrt{12}}|-\rangle_A|-\rangle_B$$

Sentence FR4: "Alice and Bob both find the outcome '−' with a probability of $\frac{1}{12}$."

The experiment is repeated n times. A contradiction arises when the four statements are combined and when, for the nth round, Bob obtains outcome "−" and knows that Alice also obtains outcome "−" (FR4). From FR1, Bob knows then that Alice's friend obtained outcome "1", and thus, from FR2, that Bob's friend obtained outcome "1". But, from FR3, this implies that Bob knows that he himself obtained outcome "+", contradicting the fact that he obtained outcome "−".

References

1. J. Szangolies, Epistemic Horizons and the Foundations of Quantum Mechanics. ArXiv e-prints, May 2018
2. S. Kochen, E. Specker, The problem of hidden variables in quantum mechanics. J. Math. Mech. **17**(1), 59–87 (1967)
3. R. Nozick, Newcomb's problem and two principles of choice, in *Essays in Honor of Carl G. Hempel*, ed. by N. Rescher (Reidel, 1969), pp. 114–146
4. S. Abramsky, S. Mansfield, R.S. Barbosa, The cohomology of non-locality and contextuality. arXiv preprint arXiv:1111.3620 (2011)
5. J.S. Bell, On the Einstein Podolsky Rosen paradox. Physics Physique Fizika **1**, 195–200 (1964)
6. S. Abramsky, R.S. Barbosa, K. Kishida, R. Lal, S. Mansfield, Contextuality, cohomology and paradox. arXiv preprint arXiv:1502.03097 (2015)
7. E.P. Wigner, Remarks on the mind-body question, in *The Scientist Speculates*, ed. by I.J. Good (Heineman, 1961)
8. D. Frauchiger, R. Renner, Quantum theory cannot consistently describe the use of itself. Nat. Commun. **9**, 3711 (2018)
9. E. Specker, Die Logik Nicht Gleichzeitig Entscheidbarer Aussagen. Dialectica **14**(2/3), 239–246 (1960)
10. M.P. Seevinck, E. specker: the logic of non-simultaneously decidable propositions (1960). arXiv preprint arXiv:1103.4537 (2011)
11. A. Peres, W.H. Zurek, Is quantum theory universally valid? Am. J. Phys. **50**(9), 807–810 (1982)
12. A.A. Abbott, C.S. Calude, K. Svozil, *On the Unpredictability of Individual Quantum Measurement Outcomes* (Springer International Publishing, Cham, 2015), pp. 69–86
13. M. Solana, *Historia de la filosofía española* (Asociacion española para el progreso de las ciencias, Madrid, 1941)
14. P. Slezak, Demons, deceivers and liars: Newcomb's malin génie. Theory Decis. **61**, 277–303 (2006)
15. A.A. Klyachko, M.A. Can, S. Binicioğlu, A.S. Shumovsky, Simple test for hidden variables in spin-1 systems. Phys. Rev. Lett. **101**, 020403 (2008)
16. L. Hardy, Nonlocality for two particles without inequalities for almost all entangled states. Phys. Rev. Lett. **71**, 1665–1668 (1993)
17. Č. Brukner, *On the Quantum Measurement Problem* (Springer International Publishing, Cham, 2017), pp. 95–117
18. A. Grinbaum, On epistemological modesty. Philosophica **83**, 139–150 (2010)
19. M. Chiara, Logical self reference, set theoretical paradoxes and the measurement problem in quantum mechanics. J. Philos. Logic **6**(1), 331–347 (1977)
20. Č. Brukner, A no-go theorem for observer-independent facts. Entropy **20**(5) (2018)
21. T. Breuer, *John von Neumann Met Kurt Gödel: Undecidable Statements in Quantum Mechanics* (Springer Netherlands, Dordrecht, 1999), pp. 159–170
22. V. Vilasini, N. Nurgalieva, L. del Rio, Multi-agent paradoxes beyond quantum theory. New J. Phys. **21**, 113028 (2019)
23. J. Bub, In defense of a 'single-world' interpretation of quantum mechanics. *Studies in History and Philosophy of Science Part B: Studies in History and Philosophy of Modern Physics* (2018)
24. S. Fortin, O. Lombardi, Wigner and his many friends: a new no-go result? (2019)
25. A. Suarez, "The limits of quantum superposition: should "Schrödinger's cat" and "Wigner's friend" be considered "miracle" narratives?" arXiv:1906.10524 preprint (2019)
26. C.A. Fuchs, On participatory realism, in *Information and Interaction* (Springer, 2017), pp. 113–134
27. J. Bub, I. Pitowsky, Two dogmas about quantum mechanics. Many Worlds 433–459 (2010)
28. Č. Brukner, A. Zeilinger, Quantum physics as a science of information, in *Quo Vadis Quantum Mechanics?* (Springer, 2005), pp. 47–61
29. C. Rovelli, Space is blue and birds fly through it. Philos. Trans. R. Soc. A: Math. Phys. Eng. Sci. **376** (2018)

30. C.A. Fuchs, B.C. Stacey, Qbism: quantum theory as a hero's handbook, in *Proceedings of the International School of Physics "Enrico Fermi"*, vol. 197 (2019), pp. 133–202
31. J.B. DeBrota, C.A. Fuchs, R. Schack, Respecting One's Fellow: QBism's Analysis of Wigner's Friend. arXiv:2008.03572 preprint (2020)
32. A. Petersen, The philosophy of Niels Bohr. Bull. At. Sci. **19**(7), 8–14 (1963)
33. H. Putnam, Is logic empirical?, in *Boston Studies in the Philosophy of Science* (Springer, 1969), pp. 216–241
34. Z. Ji, A. Natarajan, T. Vidick, J. Wright, H. Yuen, "MIP*= RE," arXiv preprint arXiv:2001.04383 (2020)
35. K. Gödel, An example of a new type of cosmological solutions of Einstein's field equations of gravitation. Rev. Mod. Phys. **21**, 447–450 (1949)
36. P. Cassou-Noguès, *Les Démons de Gödel. Logique et folie: Logique et folie* (Le Seuil, 2015)
37. A. Kull, *Self-Reference and Time According to Spencer-Brown* (Springer, Berlin, Heidelberg, 1997), pp. 71–79

Chapter 8
Epistemic Horizons: This Sentence Is $\frac{1}{\sqrt{2}}(|\text{true}\rangle + |\text{false}\rangle)$

Jochen Szangolies

Abstract In [Found. Phys. 48.12 (2018): 1669], the notion of *epistemic horizon* was introduced as an explanation for many of the puzzling features of quantum mechanics. There, it was shown that Lawvere's theorem, which forms the categorical backdrop to phenomena such as Gödelian incompleteness, Turing undecidability, Russell's paradox and others, applied to a measurement context, yields bounds on the maximum knowledge that can be obtained about a system, which produces many paradigmatically quantum phenomena. We give a brief presentation of the framework, and then demonstrate how it naturally yields Bell inequality violations. We then study the argument due to Einstein, Podolsky, and Rosen, and show how the counterfactual inference needed to conclude the incompleteness of the quantum formalism is barred by the epistemic horizon. Similarly, the paradoxes due to Hardy and Frauchiger-Renner are discussed, and found to turn on an inconsistent combination of information from incompatible contexts.

8.1 Introduction: Interpretation Versus Reconstruction

Almost from the inception of quantum mechanics, it has been clear that it does not merely represent a theory of new phenomena, but rather, an entirely novel way of theory-building. There is now wide agreement that certain assumptions and conceptions, implicit in the Newtonian, classical framework, can no longer be upheld—albeit, and perhaps shockingly, there is as yet no consensus on what, precisely, those are.

In coming to terms with the novelty of quantum mechanics, the dominant strategy has been that of *interpretation*: roughly, the attempt of matching the formalism to an underlying reality (whatever that, exactly, may mean). However, the plethora of interpretations on the market—the Wikipedia article [1] currently lists 14 'mainstream' interpretations—indicates that this project is still far from completion.

J. Szangolies (✉)
Cologne, Germany
e-mail: jochen.szangolies@gmx.de

Sometimes, the inverse of a hard problem is more easily solved. Instead of trying to infer the underlying ontology to match the quantum formalism, one might thus take a constructive road and explore which phenomena arise naturally in certain 'model' or 'toy' settings, with the aim of eventually zeroing in on QM. This is the project of *reconstructing* quantum mechanics: finding one or more foundational principles such that the quantum predictions naturally follow.

In contrast to the project of interpretation, this search has, it seems, produced a significant convergence of ideas. As pointed out by Grinbaum [2], two principles are common to several recent attempts (see references in [3]):

1. *Finiteness*: There is a finite maximum of information that can be obtained about any given system.
2. *Extensibility*: It is always possible to acquire new information about any system.

At first glance, these seem contradictory: how can we obtain additional information, if we already possess the maximum possible information about a system? The answer, as we will see, is closely related to one of the central puzzles of quantum mechanics: there must be a mechanism such that 'old' information becomes obsolete—which, in QM, is just the hotly-debated 'collapse' of the wave function.

Compare this to the situation of an observer on the spherical Earth: moving towards their horizon, bringing new terrain into view, they lose sight of what they've left behind.[1]

It may nevertheless remain mysterious why nature should conspire to withhold information from us observers. To this end, in Ref. [3], it was proposed that the principles 1 and 2 do not need to be separately postulated, but instead, follow naturally by means of applying Lawvere's fixed-point theorem [4] to the process of measurement, or more accurately, the prediction of measurement outcomes.

Lawvere's theorem essentially exposes the common (categorical) structure behind phenomena such as Gödelian incompleteness, the unsolvability of the halting problem, Russell's paradox, and many others (see [5] for an overview). Thus, by connecting it to quantum measurement, unpredictability in physics—and many quantum phenomena with it—and undecidability in mathematics can be seen as two aspects of the same phenomenon: the presence of epistemic horizons.

8.2 Horizons of Our Understanding

I do not propose to present a detailed reconstruction of the formalism of quantum mechanics here. However, I want to at least present an intuition as to how such a reconstruction, starting from the principles 1 and 2, might proceed.

---

[1] Although we typically expect that which has slipped beyond the horizon to remain largely unchanged, and thus, our information about it to remain accurate—but of course, this may not be the case.

To this end, consider as a toy model a (classical) point particle of mass m moving in one dimension. Its state can be completely described by giving its position x and velocity v—or, as is more common, its momentum $p = mv$. The space spanned by the particle's possible positions and momenta is called its *phase space*. Each point in phase space gives a tuple (x_0, p_0) uniquely determining the particle's state (see Fig. 8.1a).

From this starting point, we impose principles 1 and 2. Upon requiring that there be a maximum amount of information that can be obtained about a system, we can no longer localize its state within phase space with perfect precision—the space effectively becomes discretized (see Fig. 8.1b). Imposing then that we can always obtain additional information entails that we can increase our information about, say, its position—but to compensate, must lose information about its momentum (see Fig. 8.1c).

Thus, we do not simply obtain a discretized phase space, but rather, there is a minimum area of localization, whose shape is determined by the information obtained about each coordinate. Since position has units of length [m], while momentum has units of mass · velocity [kg$\frac{m}{s}$], this area of maximum localizabiliy has units of [kg$\frac{m^2}{s}$]—which is the dimension of Planck's famous constant, $\hbar$. Hence, maximum localizability in phase space is bounded by $\hbar$, which entails for the uncertainties Δx and Δp

$$\Delta x \Delta p \gtrsim \hbar,$$

which is of course nothing but Heisenberg's famous uncertainty relation. In this way, assumptions 1 and 2 carry us the first step of the way towards quantization.

This is, of course, an entirely heuristic picture. However, it will help, in the following, to have an intuition about the sort of project being outlined here.

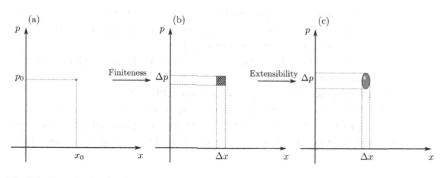

Fig. 8.1 Quantization in phase space

8.2.1 Superposition

Having now had a glimpse of how quantum phenomena emerge due to the restriction of information available about a system, it is time to consider some characteristic aspects of quantum mechanics in detail. The first step along this road will be to discuss how the impossibility of associating a definite value to every possible property of a system emerges from an argument trading on inconsistent self-reference, in much the same way as Gödelian incompletenes [6] and Turing undecidability [7].

Suppose, for simplicity, that a given system S can be in countably^2 many different states $\{s_i\}_{i\in\mathbb{N}}$—that is, there exists an enumeration $\{s_1, s_2, \ldots\}$ of states of S.

Furthermore, suppose there exists likewise an enumeration of possible measurements $\{m_j\}_{j\in\mathbb{N}}$. We will suppose that these are *dichotomic*: that is, each yields either 1 or -1 as outcome. This is not a restriction: we can always decompose a many-valued measurement into an appropriate set of dichotomic ones. Measurements are then functions that take states as input and return values, $m_n(s_k) \in \{1, -1\}$.

Think, as an example, of a coin: after we flip it, we make a measurement (that is, we look to see which side is up), and denote 'heads' as 1, 'tails' as -1. For this system, there exist only two states—s_1 for heads and s_2 for tails—and one measurement m_1, and we have

$$m_1(s_1) = 1$$
$$m_1(s_2) = -1$$

We now introduce the following assumption:

Assumption 1 (*Classicality*) For every state s_k and measurement m_n, there exists a function f such that $f(n, k) = m_n(s_k)$.

We can think of this f as a universal prediction machine for S: given the index of a state and a measurement, it spits out the result the measurement will produce. For our coin example, this function is given by Table 8.1:

Consequently, $f(1, 1) = 1$ (in state s_1, the coin shows heads), and $f(1, 2) = -1$ (in state s_2, the coin shows tails).

For the general case, with $i, j \in \mathbb{N}$, we obtain Table 8.2.

We can now lead Assumption 1 to a contradiction. To do so, we must first observe that we can construct new measurements by means of logical operations. For this, it is convenient to think of the values 1 and -1 as representing 'true' and 'false',

Table 8.1 Measurement outcomes for a coin

$f(n, k)$	s_1	s_2
m_1	1	-1

^{2} Note, however, that the argument can be generalized beyond countable sets [3].

Table 8.2 Tabulation of the function $f(n, k)$ for a general system, together with an illustration of the diagonalization technique

$f(n, k)$	s_1	s_2	s_3	s_4	s_5	$\ldots$	s_g	$\ldots$
m_1	(1)	-1	1	1	1	$\ldots$	1	$\ldots$
m_2	1	(-1)	1	-1	-1		-1	
m_3	-1	1	(-1)	-1	-1		1	
m_4	1	-1	-1	(1)	1		1	
m_5	-1	-1	-1	1	(1)		-1	
$\vdots$	$\vdots$					$\ldots$	$\vdots$	
m_g	-1	1	1	-1	-1	$\ldots$	(f)	$\ldots$
$\vdots$	$\vdots$						$\vdots$	$\ldots$

respectively. Then, we can consider $m_n(s_k) = 1$ to mean that the proposition 'S has property n in state k' is true, and $m_n(s_k) = -1$ consequently that it is false. Each measurement thus tests whether a system in a given state has or fails to have a certain property. Since properties and measurements are thus in one-to-one correspondence, we will on occasion abuse notation and speak of the 'property m_n'.

We can equivalently look at this in terms of subsets (or -regions) of the state space introduced in Fig. 8.1. Each measurement essentially tests whether the system is in some region of that space. For instance, the region with p smaller than $\sqrt{2mE_0}$ corresponds to the set of states with energy E less than E_0; a measurement that yields 1 for all states in that region (and -1 otherwise) then indicates the truth of the proposition 'S has energy less than E_0'.

This enables us to construct a logical calculus for the properties of the system. From two measurements m_1 and m_2, we can, for instance, construct $m_{12} = m_1 \oplus m_2$, where the operator $\oplus$ is taken to signify the logical xor: that is, $m_{12} = 1$ if $m_1 \neq m_2$, and $m_{12} = -1$ if $m_1 = m_2$. For ease of notation, we indicate the property values by superscripts; see Fig. 8.2.

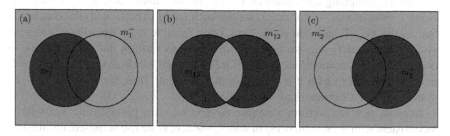

Fig. 8.2 Property-calculus in phase space

Moreover, we can give an explicit measurement procedure for each property: simply measure momentum and position up to the precision necessary to localize the state within the respective subset.

But then, this means that we can construct the following measurement m_g: for each m_i, $m_g(s_i)$ is just the opposite of $m_i(s_i)$. That is, if $m_1(s_1)$ yields 1, $m_g(s_1)$ yields -1; if $m_2(s_2)$ yields -1, then $m_g(s_2)$ yields 1. The construction of this measurement is then shown in Table 8.2.

If we now hold fast to our assumption that $f(n, k)$ enumerates all possible measurement outcomes, then m_g itself must correspond to some row of Table 8.2. However, it cannot correspond to the first row, as it differs from m_1 in the value associated to s_1; it cannot correspond to the second row, as it differs in the value associated to s_2; and so on, for any particular row of that table.

We might now suppose that, having infinitely many rows, we can just add the missing measurement. But, as is of course familiar, this move will not get us out of trouble: we can always just repeat the construction, finding a further measurement not on the list already.

But this means that there exists some state s_g and measurement m_g such that the value of $m_g(s_g)$ cannot be predicted by f. Thus, our 'universal prediction machine' cannot, in fact, exist; there are measurements such that their outcome for certain states cannot be predicted. They are, in other words, undecidable.

The above has the form of a *diagonal argument*. Diagonalization was first introduced by Cantor in his famous proof of the existence of uncountable sets, and lies at the heart of Gödel's (first) incompleteness theorem, the undecidability of the halting problem, and many others. The precise structure of such arguments in a category-theoretic setting was brought to the fore by Lawvere by means of a fixed-point theorem [4]. This can be directly adapted to the present setting, yielding a somewhat more general argument than the above; for details, see [3] and Appendix A.

An intuitive way to understand this result is the following. Consider that we can define every measurement by listing the states that lie within the corresponding subregion of state space. Then, note that we can, correspondingly, define each state via measurements—say, listing all the measurements that yield a $+1$-outcome (all the properties the system possesses in that state). Thus, we can define a measurement in terms of a state defined in terms of that very measurement—yielding the paradoxical circularity characteristic of self-reference.

This has intriguing consequences. First of all, we cannot consistently assign to s_g either a value of 1 or -1 for m_g, as supposing it ought to be 1 yields the conclusion that it must be -1, and vice versa. Thus, when faced with the question whether the system has property m_g, we find that we can neither affirm nor deny. This is, of course, just the situation Schrödinger's infamous and much-abused cat finds itself in: we can neither claim it is alive, nor that it is not. Thus, we may consider the system to be in a *superposition* with respect to m_g.

Suppose now we perform a measurement of m_g. Any possible outcome will be inconsistent with the system being in state s_g—since, as we had surmised, no outcome can consistently be associated with that state. Hence, after the measurement has yielded a result, it follows that the system can no longer be in the state s_g—that is, post measurement state change ('wave-function collapse') is a direct consequence of the preceding considerations.

It is important to note that quantum mechanics, itself, does not again fall prey to the same issues. There are two salient factors accounting for this: first, the proof depends on the possibility of 'duplicating' the index g to construct $m_g(s_g)$—which, physically, represents a cloning operation that is famously impossible in quantum mechanics [8]. Second, we must be able to invert the value of a measurement—take the value 1 to -1, and vice versa. That is, every possible value assigned to a property must be negated.

But this likewise is impossible in quantum mechanics [9]. Let us replace the classical outcomes with orthogonal quantum states $|1⟩$ and $|-1⟩$. Then, the operator

$$U_{NOT} = |1⟩⟨-1| + |-1⟩⟨1|$$

takes $|1⟩$ to $|-1⟩$, and vice versa. However, applied to the state $\frac{1}{\sqrt{2}}(|1⟩ + |-1⟩)$, we get

$$U_{NOT}\frac{1}{\sqrt{2}}(|1⟩ + |-1⟩) = \frac{1}{\sqrt{2}}(|1⟩ + |-1⟩).$$

Hence, the superposition yields a fixed point for U_{NOT}—thus evading the inconsistent assignment of Table 8.2.

This is, of course, only the first whiff of quantum phenomena. The picture can be developed further. Complementarity, the impossibility to simultaneously assign definite values to certain properties, can be obtained by considering a form of the above argument in the context of sequences of measurements. Furthermore, the uncertainty principle emerges as a finite bound on the information available about a system—the number of simultaneously definite properties—by appealing to Chaitin's version of the incompleteness theorem [10]. For details, see [3]. In the following, we will consider another paradigmatically quantum feature that has, so far, not been considered: entanglement.

For this, it will be useful to consider a simple 'toy' system. Thus, take the extreme case of a system S such that only one of its properties is decidable—all save a single bit of information lies beyond the epistemic horizon. We are then in the situation of Fig. 8.2: one bit of information decides one of three possible mutually exclusive measurements on the system. Appreciating the parallel to the orthogonal measurements for a single qubit, we will name these three properties x_S, y_S, and z_S.

8.2.2   Entanglement

So far, we have only considered single, individual systems. One might therefore ask what this framework entails once one investigates composite systems instead. Thus, take two systems, A and B. To keep matters simple, we will confine our discussion here to 'toy systems' of the kind introduced above—that is, systems described by a single definite property.

Consequently, $\mathcal{A}$ is described by either of x_A, y_A, or z_A having a definite value, while $\mathcal{B}$'s state is given by one out of x_B, y_B, and z_B. A possible state of the compound system $\mathcal{A} \otimes \mathcal{B}$ would then be (x_A^+, x_B^-), where we used the superscript notation to indicate property values.

As we had surmised, however, we can use elementary Boolean logic to construct new properties. Thus, let us consider the property $x_{AB} = x_A \oplus x_B$. This indicates a correlation between the two x-values: it signifies that one must be the opposite of the other. We can then e.g. give a complete description of the system as (x_A^+, x_{AB}^+), signifying the state where $x_A = 1$, and x_B must be the opposite, hence -1.

This is, as yet, a completely classical situation. Picture the case of two colored cards, one red, and one green, in two envelopes: once you open one, you immediately know the color of the card within the other, even if the latter is located on Pluto. There is in particular nothing nonlocal about you having this knowledge.

However, consider now the state (x_{AB}^+, z_{AB}^+). Here, the two bits of information we have available to describe the state are entirely taken up by the correlations: we know that the two x-values, as well as the two z-values, are opposed to one another; but we know nothing whatever about any individual x- or z-value!

This is precisely the situation of an entangled two-particle system (cf. [11]). Our $f(n, k)$, which, for this system, can only determine two properties, only provides values for x_{AB} and z_{AB}, but leaves, e.g., z_B undecidable. However, once we have performed the requisite measurement, the considerations of the previous sections tell us that something remarkable must happen: whatever outcome is produced, one bit of information must now be taken up by the value of z_B; but, due to the (anti-)correlation between z-values, this then immediately tells us the value of z_A, as well! Furthermore, as all information available is now taken up by (e.g.) (z_B^+, z_{AB}^+), it follows that nothing about the x-values can be known: the correlation there is destroyed.

8.3 Does This Ring a Bell?

We now have the tools in hand to investigate one of the most famous expressions of quantum 'weirdness': Bell's theorem [12], or the failure of 'local realism'. We will start with a slightly different view on Bell inequalities [13].

Consider, to this end, again that there exists a function $f(n, k)$ providing values to all possible measurements. In particular, consider the above bipartite system and the properties x_A, z_A, x_B and z_B. With respect to these properties, every state can be written as a four-tuple (x_A, z_A, x_B, z_B), corresponding to a column in Table 8.2. That is, there are 16 possible states, from $f(n, 1) = (x_A^+, z_A^+, x_B^+, z_B^+)$ to $f(n, 16) = (x_A^-, z_A^-, x_B^-, z_B^-)$, which we label λ_i.

In any given experiment, each of these states may be present with a certain probability $P(\lambda_i) = p_i$. See Table 8.3 for an enumeration.

Table 8.3 States and probabilities for the bipartite syste $\mathcal{A} \otimes \mathcal{B}$

State	$x_{\mathcal{A}}$	$z_{\mathcal{A}}$	$x_{\mathcal{B}}$	$z_{\mathcal{B}}$	$P(\lambda_i)$
λ_1	1	1	1	1	p_1
λ_2	1	1	1	-1	p_2
λ_3	1	1	-1	1	p_3
λ_4	1	1	-1	-1	p_4
λ_5	1	-1	1	1	p_5
λ_6	1	-1	1	-1	p_6
λ_7	1	-1	-1	1	p_7
λ_8	1	-1	-1	-1	p_8
λ_9	-1	1	1	1	p_9
λ_{10}	-1	1	1	-1	p_{10}
λ_{11}	-1	1	-1	1	p_{11}
λ_{12}	-1	1	-1	-1	p_{12}
λ_{13}	-1	-1	1	1	p_{13}
λ_{14}	-1	-1	1	-1	p_{14}
λ_{15}	-1	-1	-1	1	p_{15}
λ_{16}	-1	-1	-1	-1	p_{16}

With this, we can compute probabilities for individual outcomes by marginalization—that is, summing over all probabilities for states that contain the desired outcome. Therefore, the probability to find $x_{\mathcal{A}} = 1$ is equal to $P(x_{\mathcal{A}}^+) = \sum_{i=1}^{8} p_i = p_1 + p_2 + \cdots + p_8$, as states λ_1 through λ_8 have $x_{\mathcal{A}} = 1$. We can likewise compute probabilities for joint events: $P(x_{\mathcal{A}}^+, x_{\mathcal{B}}^-) = p_3 + p_4 + p_7 + p_8$.

Finally, we can compute expectation values for such joint events:

$$\langle x_{\mathcal{A}} x_{\mathcal{B}} \rangle = \sum_{r,s \in \{1,-1\}} rs P(x_{\mathcal{A}}^r, x_{\mathcal{B}}^s)$$

$$= P(x_{\mathcal{A}}^+, x_{\mathcal{B}}^+) + P(x_{\mathcal{A}}^-, x_{\mathcal{B}}^-) - P(x_{\mathcal{A}}^+, x_{\mathcal{B}}^-) - P(x_{\mathcal{A}}^-, x_{\mathcal{B}}^+)$$

$$= p_1 + p_2 - p_3 - p_4 + p_5 + p_6 - p_7 - p_8 - p_9 - p_{10} + p_{11} + p_{12} - p_{13} - p_{14} + p_{15} + p_{16}$$

These expectation values carry information about the correlation between the two properties: if $\langle x_{\mathcal{A}} x_{\mathcal{B}} \rangle = 1$, only the p_i with a positive sign are nonzero, and thus, $x_{\mathcal{A}} = x_{\mathcal{B}}$ for all states in the ensemble; for $\langle x_{\mathcal{A}} x_{\mathcal{B}} \rangle = -1$, we obtain $x_{\mathcal{A}} = -x_{\mathcal{B}}$. If $\langle x_{\mathcal{A}} x_{\mathcal{B}} \rangle = 0$, the value of $x_{\mathcal{A}}$ tells us nothing about $x_{\mathcal{B}}$, and vice versa.

With this, it is easy to compute the quantity

$$\langle C_{CHSH} \rangle = \langle x_{\mathcal{A}} x_{\mathcal{B}} \rangle + \langle x_{\mathcal{A}} z_{\mathcal{B}} \rangle + \langle z_{\mathcal{A}} x_{\mathcal{B}} \rangle - \langle z_{\mathcal{A}} z_{\mathcal{B}} \rangle$$

$$= 2 - 4(p_3 + p_4 + p_6 + p_8 + p_9 + p_{11} + p_{13} + p_{14})$$

$$= 4(p_1 + p_2 + p_5 + p_7 + p_{10} + p_{12} + p_{15} + p_{16}) - 2.$$

Since $\sum_i p_i \leq 1$, this immediately yields

$$-2 \leq \langle C_{CHSH} \rangle \leq 2.$$

This is, of course, nothing but the famous CHSH-Bell inequality [14].

This should strike us as somewhat remarkable: only the assumption that there exists a $f(n, k)$ assigning values to all observables turns out to be enough to derive a bound on the above expression. Thus, Bell inequalities precisely delineate the set of theories for which there exists $f(n, k)$ such that it yields the values for all possible measurements. Contrariwise, Bell inequality violations certify that no such $f(n, k)$ for all values can exist—or at least, be probed by experiment.

The undecidability of these values then allows for the violation of this expression—as is, indeed, observed in quantum mechanics. Mathematically, the bounds on $\langle C_{CHSH} \rangle$ correspond to necessary conditions for the existence of a joint probability distribution (Table 8.3); their violation means that no consistent assignment of probabilities to the λ_i is possible.

This should not surprise us: we have already seen that, for instance, the event $(x_{\mathcal{A}}^+, z_{\mathcal{A}}^+)$ cannot occur—$f(n, k)$ does not assign simultaneous values to both elements. But if the above probability distribution were to exist, we could easily obtain

$$P(x_{\mathcal{A}}^+, z_{\mathcal{A}}^+) = p_1 + p_2 + p_3 + p_4.$$

But what could it mean to assign a probability to an impossible event?

It is more usual to attribute violations of Bell inequalities to the failure of either *locality* or *realism*. What does the above probability distribution have to do with either?

'Realism' is ultimately simply the possibility of assigning values to all observables. If such an assignment is possible, each of the rows in Table 8.3 designates a valid state, and can be assigned a probability, leading to the above considerations.

But how is the failure of locality supposed to avoid this trouble? The resolution here is that we have implicitly assumed that we can fairly sample from the above probability distribution. However, if outcome probabilities on $\mathcal{B}$ were to change due to measurements on $\mathcal{A}$, then we could no longer carry the argument through. Hence, one usually makes an assumption that a measurement on one part of the system does not influence measurements carried out on the other; to make this assumption sensible, one ensures that both parts of the system are far away from one another, such that no influence, propagating at the speed of light, could travel between them. Should there then be any instantaneous influence despite these precautions, we speak of a failure of locality.

8.4   EPistemic HoRizons: Incomplete Quantum Mechanics?

It is sometimes proposed that Bell's theorem only hinges on the assumption of locality, and hence, its violation suffices to conclude that nature is nonlocal (e.g. [15]). The reasoning here is typically that 'realism' is not a separate requirement that could fail on its own, but rather, is already established by the famous argument due to Einstein, Podolski, and Rosen (EPR) [16].

Let us take a lightning-quick review of the argument adapted to the present formalism. EPR take a system in the state (x_{AB}^+, z_{AB}^+), and consider measurements on one of its parts (say $\mathcal{A}$). Upon measuring $x_{\mathcal{A}}$, we obtain the x-value for $\mathcal{A}$, and due to the correlation given by x_{AB}^+, can immediately infer $x_{\mathcal{B}}$; likewise for z. However, the quantum formalism does not permit us to speak of simultaneous values for $x_{\mathcal{B}}$ and $z_{\mathcal{B}}$. But how, then, is $\mathcal{B}$ supposed to know to 'produce' the right value in each case?

The EPR-argument hinges on a bit of *counterfactual reasoning*: *had* we measured $z_{\mathcal{A}}$ (instead of $x_{\mathcal{A}}$), we *would have* been able to predict a definite value for $z_{\mathcal{B}}$ (instead of $x_{\mathcal{B}}$). Due to the absence of any disturbance on $\mathcal{B}$ due to our actions on $\mathcal{A}$ (locality), we then conclude that $\mathcal{B}$ cannot just spontaneously 'decide' which value to produce, and hence, both $x_{\mathcal{B}}$ and $z_{\mathcal{B}}$ must have had a definite value—in EPR's parlance, an 'element of reality'—associated to them all along.

To illustrate this puzzle, Schrödinger introduced the analogy of the fatigued student [17]: quizzed in an oral examination, they will get the first answer right with certainty, after which, however, any further answer will be random. Even though we only get one correct answer out in any case, we still must conclude that the student knew the answer to every question, in order to produce this performance: *had* we asked a different first question, then nevertheless the student *would have* produced the right answer.

Applied to quantum mechanics, this would entail that the description of the correlated system $\mathcal{A} \otimes \mathcal{B}$ must be incomplete: $\mathcal{B}$ must, to give the right answer in each of these cases, 'know' the correct values for $x_{\mathcal{B}}$ and $z_{\mathcal{B}}$ in advance, these answers simply being hidden to the quantum formalism.

If this is correct, then nonlocality is our only out in the case of Bell's theorem: there are simultaneous values for all observables—$f(n, k)$ does not tell the whole story—and measuring one part of a system must influence the value distribution of the distant part.

One way to attempt to defuse the force of EPR's argument is to deny that the sort of counterfactual inference that allows us to reason about what would have happened had our measurement choice been different is valid, at least in a quantum context. However, without further substantiation regarding why that should be the case, simply denying the validity of a certain form of argument to avoid an unwelcome conclusion hardly seems fair.

While I do not presume to settle this controversy once and for all, I believe the present framework offers a fresh perspective on the matter. For consider what happens in each of the two cases. The initial state (x_{AB}^+, z_{AB}^+) becomes, say, $(z_{\mathcal{A}}^+, z_{AB}^+)$, respectively $(x_\mathcal{A}^+, x_{\mathcal{AB}}^+)$. The only change is thus in the properties of the local system $\mathcal{A}$, about which we have gained new information.

This allows us then to *infer* the value of the distant system. However, we may hold that this is something different than that value spontaneously becoming definite— after all, this value is not given by any $f(n, k)$. We could thus associate 'elements of reality' to the values of $f(n, k)$ exclusively. Our conclusions about the distant values would then have the status of inferences about the truth value of the Gödel sentence: We can *infer* that 'I am not provable (in a given axiomatic system)' is true, since it is, in fact, not provable (in that system); however, the system itself will not be able to establish this truth (on pain of contradiction).

Such a state is one in which we have the following two items of knowledge: 'the x/z-value of $\mathcal{A}$ is 1' and 'the x/z-value of $\mathcal{B}$ is opposite that of $\mathcal{A}$'. This differs from a state like $(x_\mathcal{A}^+, x_\mathcal{B}^-)$ in a subtle, but crucial, way. In that state, our knowledge is given by 'the x-value of $\mathcal{A}$ is 1' and 'the x-value of $\mathcal{B}$ is -1'. The difference emerges if we imagine varying the first of each set of propositions—that is, engage in counterfactual reasoning. In case of a state like $(x_\mathcal{A}^+, x_\mathcal{B}^-)$, we can say that *had* we obtained a value of 1 for the z-value instead, we could still validly speak of the x-value of $\mathcal{B}$ being -1.

That is not the case for the state $(x_\mathcal{A}^+, x_{\mathcal{AB}}^+)$: varying the first proposition, but leaving the second constant, would lead us to a state in which we have *no information* about the x-value of $\mathcal{B}$. Consequently, the two states differ in the counterfactuals they support: the state $(x_\mathcal{A}^+, x_\mathcal{B}^-)$ allows us to say that, *had* the first value been different, the second *would have been the same* (absent any disturbance), leading to e.g. $(z_\mathcal{A}^+, x_\mathcal{B}^-)$. However, in the state $(x_\mathcal{A}^+, x_{\mathcal{AB}}^+)$, as soon as we imagine exchanging $x_\mathcal{A}$, we loose any ability to make determinations of $x_\mathcal{B}$, as this value is specified only contingently on that of $x_\mathcal{A}$. In a state like '$(z_\mathcal{A}^+, x_{\mathcal{AB}}^+)$', $x_\mathcal{B}$ would simply not have any determinate value at all.

An alternative way to think about the situation is by introducing the notion of a *conditional event*. A conditional event is, for instance, an observable that only takes a value conditionally on the value of another. Thus, we can write the state $(x_\mathcal{A}^+, x_{\mathcal{AB}}^+)$ equivalently as $(x_\mathcal{A}^+, x_\mathcal{B}^-|x_\mathcal{A}^+)$, where the '|'-notation denotes the conditioning: the value of $x_\mathcal{B}$ is -1 *given that* the value of $x_\mathcal{A}$ is $+1$. This is a rewriting of the information contained in $x_{\mathcal{AB}}^+$ that more clearly emphasizes the result of $\mathcal{A}$ observing a given value on their ability to predict the value of $\mathcal{B}$'s measurement.

We should then not think about the state $(x_\mathcal{A}^+, x_\mathcal{B}^-|x_\mathcal{A}^+)$ as containing the value of $\mathcal{B}$'s measurement; rather, the two bits of information it contains jointly entail $\mathcal{B}$'s value. This is a salient difference. Consider, for example, the case of a one-time pad: one bit of the encrypted message plus one bit of the key may entail one bit of the clear text, in the same way that $\mathcal{A}$'s measurement result, plus knowledge of the correlation, entails $\mathcal{B}$'s outcome. But that does not mean that the state, as such, must contain information about $\mathcal{B}$'s value if $\mathcal{A}$'s value were different, anymore than the value of the key alone must contain information about the clear text.

Thus, only given that one has actually measured $x_\mathcal{A}$ is reasoning about the value of $x_\mathcal{B}$ possible. In this sense, the present framework gives a natural meaning to Bohr's somewhat opaque 'influence on the precise conditions which define the possible types of prediction which regard the subsequent behaviour of the system' [18]. We naturally imagine it to be possible to change one thing, while keeping something else equal; but in this case, the 'one thing' (the definite value of x_A) is part of the antecedent conditions for making determinations about that 'something else' (the value of x_B). Moreover, trying to simultaneously appeal to both $(x_A^+, x_B^- | x_A^+)$ and $(z_A^+, z_B^- | z_A^+)$ (for instance) amounts to exceeding the information bound on the system as given by $f(n, k)$; thus, the illegitimate nature of the counterfactual argument in this case is seen to be rooted in the more fundamental informational limit. The epistemic horizon puts a limit to the information accessible about any given system, and each attempt to access more courts inconsistency.

The EPR argument, then, essentially trades on a conflation of $(x_A^+, x_B^- | x_A^+)$ with (x_A^+, x_B^-). Only the latter state supports the reasoning that leads us to conclude that the distant particle must have 'known' the value of both x_B and z_B all along.

8.5 Hardy's Paradox

The tools developed above can be fruitfully applied to other supposed 'paradoxes' in the quantum world. Consider, to this end, the Hardy state [19, 20] of two entangled qubits, which is in the $z_A z_B$-basis

$$|\psi_H\rangle = \frac{1}{\sqrt{3}} \left(|z_A^+ z_B^+\rangle + |z_A^+ z_B^-\rangle + |z_A^- z_B^+\rangle \right). \tag{8.1}$$

Here, a state such as $|z_A^+ z_B^+\rangle$ means that A and B would obtain the values z_A^+ resp. z_B^+ upon performing the requisite measurements.

Hardy's paradox now consists in pointing out that elementary reasoning suffices to demonstrate that two parties, A and B, measuring each qubit in the x-basis $\{|x^+\rangle, |x^-\rangle\} = \{\frac{1}{\sqrt{2}}(|z^+\rangle + |z^-\rangle), \frac{1}{\sqrt{2}}(|z^+\rangle - |z^-\rangle)\}$, can never both see the state $|x^-\rangle$ (i.e. obtain the outcomes x_A^- and x_B^-). Yet, in fact, this happens with a probability $p_H = \frac{1}{12}$.

This can be seen by writing $|\psi_H\rangle$ in the $x_A x_B$-basis. This yields:

$$|\psi_H\rangle = \frac{3}{\sqrt{12}} |x_A^+ x_B^+\rangle + \frac{1}{\sqrt{12}} |x_A^+ x_B^-\rangle + \frac{1}{\sqrt{12}} |x_A^- x_B^+\rangle - \frac{1}{\sqrt{12}} |x_A^- x_B^-\rangle \tag{8.2}$$

According to the Born rule, measurements performed on this state yield the $|x_A^- x_B^-\rangle$-outcome with probability $p_H = \frac{1}{12}$.

It is useful, here, to look at the chain of reasoning used to arrive at the above conclusion in greater detail (cf. [21]). To start with, in the above notation, for the Hardy state in the $z_A z_B$-basis, the state contains the information that 'if B obtains the outcome z_B^-, then A obtains the outcome z_A^+'; once B then obtains that outcome, we have the information content $(z_B^-, z_A^+ | z_B^-)$.

As before, the notation '$z_A^+|z_B^-$' expresses the conditional nature of $\mathcal{A}$'s value; only *given that* $\mathcal{B}$ obtained the value z_B^- can we consistently talk about $\mathcal{A}$'s observed value.

Now, the state in the $x_A z_B$-basis is:

$$|\psi_H\rangle = \sqrt{\frac{2}{3}}\,|x_A^+ z_B^+\rangle + \frac{1}{\sqrt{6}}\,|x_A^+ z_B^-\rangle + \frac{1}{\sqrt{6}}\,|x_A^- z_B^-\rangle \tag{8.3}$$

From this, we see that, if $\mathcal{A}$ measures $x_A = -1$, $\mathcal{B}$ must obtain $z_B = -1$, that is, the information within the state afterwards is $(x_A^-, z_B^-|x_A^-)$.

Finally, in the $z_A x_B$-basis, the state is:

$$|\psi_H\rangle = \sqrt{\frac{2}{3}}\,|z_A^+ x_B^+\rangle + \frac{1}{\sqrt{6}}\,|z_A^- x_B^+\rangle + \frac{1}{\sqrt{6}}\,|z_A^- x_B^-\rangle \tag{8.4}$$

If $\mathcal{A}$ thus obtains $z_A = +1$, $\mathcal{B}$ must obtain $x_B = +1$, and the information content afterwards is $(z_A^+, x_B^+|z_A^+)$.

This now suffices to establish the contradiction. Suppose we were to reason as follows:

(i) If $\mathcal{A}$ obtains x_A^-, we can conclude that $\mathcal{B}$ must obtain z_B^-, due to 3.
(ii) Thus, suppose $\mathcal{B}$ then in fact obtains z_B^-. With 1, we can then conclude that $\mathcal{A}$, had she measured in the z_A-basis, would obtain z_A^+.
(iii) However, if $\mathcal{A}$ obtains z_A^+, then 4 tells us that $\mathcal{B}$ must obtain x_B^+.
(iv) (From i–iii) Putting these together, we surmise that if $\mathcal{A}$ obtains x_A^-, $\mathcal{B}$ must obtain x_B^+, and consequently, the outcome $|x_A^- x_B^-\rangle$ can never occur.
(v) Yet, by 2, $|x_A^- x_B^-\rangle$ occurs with probability $p = \frac{1}{12}$. ⚡

To see what goes wrong here, let us go back, for a moment, to the discussion of the EPR paradox. There, we surmised that the information within a state $(x_A^+, x_B^-|x_A^+)$ crucially differs from that in a state like (x_A^+, x_B^-) in that the latter, but not the former, supports counterfactual inferences. That is, if we have the information about both systems individually, we can imagine varying the value of one system independently; but if the information about one system is only specified conditionally on that of the other, then counterfactual reasoning becomes nonsensical.

The EPR argument would successfully establish the incompleteness of quantum mechanics if, when $\mathcal{A}$ measures in the x-basis, we had the state (x_A^+, x_B^-), and likewise, for z-measurement, the state (z_A^+, z_B^-). For then, we could say that if $\mathcal{A}$ *had measured* in a basis different from the one in which she actually did measure in any given experiment, $\mathcal{B}$'s particle nevertheless must have been prepared to produce a fitting answer. These two states could hence be termed counterfactually consistent, and we can appeal to both in a single argument.

However, that is not the case for states of the form $(x_A^+, x_B^-|x_A^+)$. Here, $\mathcal{B}$'s value is only specified conditional on $\mathcal{A}$'s; thus, we cannot consistently imagine varying only $\mathcal{A}$'s value, as it forms part of the determining conditions of $\mathcal{B}$'s value. The state

$(x_A^+, x_B^- | x_A^+)$ and its counterpart $(z_A^+, z_B^- | z_A^+)$ are thus not counterfactually consistent, and cannot be used in a single argument.

But the same is then true for $(x_A^-, z_B^+ | x_A^-)$ and $(z_A^+, x_B^+ | z_A^+)$. Both apply only in the contexts in which $\mathcal{A}$ did, in fact, make the x- respectively z-basis measurement. Since $\mathcal{A}$ cannot in fact make both measurements, propositions (i) and (iii) cannot simultaneously be appealed to: their combination would exceed the amount of information consistently obtainable about the system.

Consequently, the 'paradox' in the above argument is of the same nature as that due to EPR, and similarly tells us that, in a quantum world, we must be careful which propositions about a system are simultaneously definite, and thus, can be used to underwrite counterfactual arguments.

8.6 The Frauchiger-Renner Argument

Recently, an intriguing new argument has been presented by Daniela Frauchiger and Renato Renner [22]. They aim to show that "quantum theory cannot consistently describe the use of itself", and use an ingenious thought experiment to support their claim. The paper has already received much commentary, which points both to the high impact and controversial nature of their result, as well as to the lack of consensus regarding its interpretation.

The Frauchiger-Renner argument can be read as a 'Wigner's Friendification' [23] of Hardy's paradox. In a famous Gedankenexperiment [24], Wigner ($\mathcal{W}$) asks us to imagine a hermetically sealed laboratory containing a scientist (the eponymous 'Friend' $\mathcal{F}$) carrying out a Schrödinger's cat-type experiment. At some point, $\mathcal{F}$ will have made some definite observation of the cat's well-being. Yet, $\mathcal{W}$, having no knowledge of $\mathcal{F}$'s result (although he may have knowledge that $\mathcal{F}$ has observed *some* definite result), must, applying the usual rules of quantum mechanics, describe the entire laboratory system as being in a state of superposition. Indeed, in theory, he could perform an interference experiment on the entire laboratory that would confirm his description.

But this poses a problem: $\mathcal{F}$, we should expect, has made a definite observation, yet $\mathcal{W}$'s description and experimental results are incompatible with any given definite state of the laboratory system.

The 'Wigner's Friend'-scenario is essentially a 'Wigner's Friendification' of the EPR-argument: the latter features two entangled systems, while the former makes one of these systems a conscious observer, and adds another observer (a 'meta-observer', [21]) which carries out a measurement on the total system in an orthogonal basis. $\mathcal{W}$, we imagine, knows that the system is either in the state (cat alive, friend sees cat alive) or (cat dead, friend sees cat dead)—since both are incompatible with interference, we conclude, there must be some contradiction. Perhaps $\mathcal{F}$'s observation collapses the wave function, and thus, standard quantum rules no longer obtain once conscious observation is involved.

However, crucially, according to the above discussion, $\mathcal{W}$ in fact only knows that the system is in the state (cat alive, friend sees cat alive|cat alive) or (cat dead, friend sees cat dead|cat dead). And these, we had surmised, cannot be simultaneously appealed to consistently. Hence, the conclusion of a contradiction does not, in fact, obtain.

Frauchiger and Renner now essentially formulate a Wigner's-Friendified version of the Hardy paradox: consider two observers, $\mathcal{A}$'s friend $\mathcal{F}_A$ and $\mathcal{B}$'s friend $\mathcal{F}_B$, which share an entangled two-qubit system, and perform z-basis measurements on their respective qubits. In the state

$$|\psi_H\rangle = \frac{1}{\sqrt{3}}\left(|z_A^+ z_B^+\rangle + |z_A^+ z_B^-\rangle + |z_A^- z_B^+\rangle\right),$$

we now consider, e.g., z_A^+ to be the 'belief state' of $\mathcal{A}$'s friend $\mathcal{F}_A$ after performing a z-measurement and obtaining the outcome $+1$—analogous to $\mathcal{F}$'s state after observing the cat. $\mathcal{A}$ and $\mathcal{B}$ then carry out their measurements on the entire laboratories containing their respective friends in the basis $\{|x^+\rangle, |x^-\rangle\} = \{\frac{1}{\sqrt{2}}(|z^+\rangle + |z^-\rangle), \frac{1}{\sqrt{2}}(|z^+\rangle - |z^-\rangle)\}$, as before. However, this is now to be interpreted as a measurement testing for the superposed states of the entire laboratories, containing their respective friends, here labeled by their respective 'belief states'.

As before, simple application of the Born rule immediately tells us that both $\mathcal{A}$ and $\mathcal{B}$ may observe the -1-outcome with probability $\frac{1}{12}$. We can now again apply the reasoning of Hardy's paradox to obtain the apparent contradiction. However, in this version, the argument has an added wrinkle: we are not merely thinking about results $\mathcal{A}$ (say) *would have* obtained, *had* she made the appropriate measurements, but about measurements actually performed by $\mathcal{F}_A$. Does this change matters?

From Eq. (8.3), we find that, if $\mathcal{A}$ obtains -1, $\mathcal{F}_B$ must obtain -1, likewise. But then, if $\mathcal{F}_B$ obtains -1, Eq. (8.1) tells us that $\mathcal{F}_A$ must obtain the $+1$-outcome. Finally, Eq. (8.4) tells us that given that $\mathcal{F}_A$ sees $+1$, $\mathcal{B}$ must obtain the $+1$-outcome.

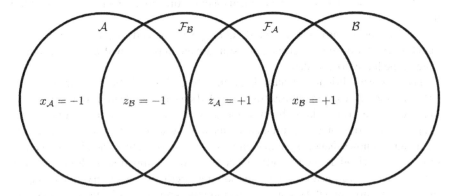

Fig. 8.3 Epistemic horizons of the observers in the Frauchiger-Renner argument

In summary: having obtained the value -1 in her measurement, $\mathcal{A}$ knows that $\mathcal{F}_B$ knows that $\mathcal{F}_A$ knows that B must obtain the value $+1$, and thus, knows herself that B must obtain the value $+1$; yet, with probability $\frac{1}{12}$, both $\mathcal{A}$ and B obtain the outcome -1.

The point of the Wigner's-Friendification is then the following: we are now not considering different measurements that $\mathcal{A}$ *could have* performed (but didn't), but rather, measurements as actually performed by distinct observers, who presumably have each obtained definite measurement results. It is then tempting to think of these as 'facts in the world', available for classical—that is, Boolean—logical reasoning.

But there still is no unified logical framework encompassing both $(x_{\mathcal{A}}^-, z_B^- | x_{\mathcal{A}}^-)$ and $(z_{\mathcal{A}}^+, x_B^+ | z_{\mathcal{A}}^+)$—appealing to both simultaneously amounts to exceeding the information limitation about the system as given by $f(n, k)$. $\mathcal{F}_B$'s observation of -1 can be contained within $\mathcal{A}$'s epistemic horizon, but $\mathcal{F}_B$'s determination of $\mathcal{F}_A$'s observation cannot also be: as shown in Fig. 8.3, each individual epistemic horizon contains at most two bits of information. The Frauchiger-Renner argument then exceeds that limit by trying to unite the different, overlapping horizons into one—an impossibility already highlighted by the impossibility of finding a joint probability distribution over all observables in a Bell experiment. Hence, the 'telescoping' of knowledge necessary for $\mathcal{A}$'s conclusion that B can never see -1 if she sees -1 cannot be performed: $\mathcal{A}$'s attempt to peek behind her epistemic horizon fails.

Frauchiger and Renner codify this 'telescoping' in their assumption C, which says that "a theory T must [...] allow any agent A to promote the conclusions drawn by another agent A' to his own conclusions" [22]. This assumption, then, fails to be satisfied, if the preceding framework is apt. However, this is not an instance of quantum theory failing to "consistently describe the use of itself"; rather, quantum theory, as already established by EPR and Hardy-type arguments, restricts which propositions can be consistently combined, without exceeding the bound on the maximal information that can be contained within a system.

One important lesson of the Frauchiger-Renner argument then is that it is not the counterfactual reasoning, per se, that is problematic in quantum mechanics, but rather, exceeding the informational limitation given by the undecidable values of $f(n, k)$. In the EPR and Hardy-arguments, this limitation is exceeded via counterfactually appealing to values that would have been obtained, had different measurements been carried out; but even if, as in the case of the FR argument, these measurements are actually performed, the bound on the maximum information available for any given system prohibits appealing to them within a single argumentative context.

8.7 Conclusion

We have considered the application of self-referential arguments to physical systems, and found that many paradigmatically quantum phenomena seem to gain a natural explanation from this perspective. This idea is not entirely new: John Wheeler himself proposed the undecidable propositions of mathematical logic as a candidate for a 'quantum principle', from which to derive the phenomenology of quantum mechanics [25]—a proposal which, as legend has it, got him thrown out of Gödel's office [26]. For a brief review of these efforts, see [3] and references therein.

What this program, if successful, shows is that there is a common thread behind mathematical undecidability and physical unknowability—that, in other words, the epistemic horizons the pure mathematician and the experimental physicist find delimiting their perspectives are not separated, but instead, spring from a common source.

In an intriguing sense, the incompleteness of mathematics may then come to the rescue of physics, allowing it in turn to yield a complete picture: the incompleteness the EPR-argument seeks to establish is averted by the horizon that bars counterfactual reasoning about unperformed experiments—which, hence, famously 'have no results' [27]. It is as if Schrödinger's student does not know the answer to any questions, as such, but knows each answer only relative to that question being asked.

This motivates a proposal of *relative realism*: assign 'elements of reality' only where $f(n, k)$ yields a definite value. In this way, we get as close to the classical ideal of local realism as is possible in a quantum world. The resolution of the EPR, Hardy, and Frauchiger-Renninger paradoxes is then to deny the EPR notion of 'elements of reality': according to their definition, an element of reality is associated with every value that can be predicted with certainty. But in a state such as $(x_A^+, x_B^- | x_A^+)$, we can predict x_B^- with certainty, but no element of reality is associated to it; rather, it is the conditional value $x_B^- | x_A^+$ that is definite in this sense, which does not allow us to make any determination of x_B in the absence of a definite value for x_A, and which cannot stand for x_B^- in chains of inferences.

We may try, combining indirectly-obtained information from different contexts in ever more ingenious ways, to look beyond our epistemic horizon; but the Old One's secrets, it seems, are not so easily discerned.

Technical Endnotes

A. The Lawvere Fixed-Point Argument

We will explicitly construct a measurement $m_g(s_k)$, that is, a function $m_g : \Sigma_S \to \{1, -1\}$, where Σ_S denotes the state space of S, such that it differs from $f(n, k)$ for at least one s_k.

Suppose that there exists a function $f(n, k) : \mathbb{N} \times \mathbb{N} \to \{1, -1\}$ such that it is equal to the outcome of the nth measurement for the kth state. Furthermore, we introduce the arbitrary map $\alpha : \{1, -1\} \to \{1, -1\}$, and the map $\Delta : \mathbb{N} \to \mathbb{N} \times \mathbb{N}$ that takes $n \in \mathbb{N}$ to the tuple $(n, n) \in \mathbb{N} \times \mathbb{N}$. With these, we construct g as the map that makes the following diagram commute:

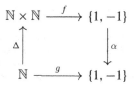

The map g constructed in this way then yields sequentially values for a certain measurement, m_g, if performed on states of $\mathcal{S}$, i.e. $g(k) = m_g(s_k)$. If f yields the value of every measurement applied to every state, then there must be some n such that $g(k) = f(n, k)$ for all states s_k. Choose now $k = n$ and evaluate $g(n)$:

$$f(n, n) = g(n)$$
$$= \alpha(f(n, n))$$

The first equality is simply our stipulation that g should encode some measurement, and that $f(n, n)$ yields the outcome of the nth measurement on the nth state. The above then shows that the map α must have a fixed point at $f(n, n)$ for the construction to be consistent.

However, we are free in our choice of α, and consequently, may choose the negation $\neg(1) = -1, \neg(-1) = 1$. But this clearly has no fixed point, and we obtain the contradiction

$$f(n, n) = \neg f(n, n) \quad \lightning$$

But then, this means that no f reproducing every measurement outcome can exist.

References

1. Wikipedia contributors, *Interpretations of quantum mechanics | Wikipedia, The Free Encyclopedia.* [Online; accessed 17-January-2020] (2019), https://en.wikipedia.org/w/index.php?title=Interpretations_of_quantum_mechanics&oldid=930824996
2. A. Grinbaum, Elements of information-theoretic derivation of the formalism of quantum theory. Int. J. Quantum Inf. **1**(03), 289–300 (2003)
3. J. Szangolies, Epistemic horizons and the foundations of quantum mechanics. Found. Phys. **48**(12), 1669–1697 (2018)
4. F.W. Lawvere, Diagonal arguments and cartesian closed categories, in *Category Theory, Homology Theory and Their Applications II* (Springer, 1969), pp. 134–145
5. N.S. Yanofsky, A universal approach to self-referential paradoxes, incompleteness and points. Bull. Symb. Logic **9**(03), 362–386 (2003)
6. K. Göodel, Über formal unentscheidbare Sätze der Principia Mathematica und verwandter Systeme I. Monatshefte füur Mathematik und Physik **38**(1), 173–198 (1931)
7. A.M. Turing, On computable numbers, with an application to the Entscheidungsproblem. J. Math. **58**(345-363), 5 (1936)

8. W.K. Wootters, W.H. Zurek, A single quantum cannot be cloned. Nature **299**(5886), 802–803 (1982)
9. K. Svozil, A constructivist manifesto for the physical sciences—Constructive re-interpretation of physical undecidability, in *The Foundational Debate* (Springer, Dordrecht, 1995), pp. 65–88
10. G.J. Chaitin, Information-theoretic incompleteness. Appl. Math. Comput. **52**(1), 83–101 (1992)
11. A. Zeilinger, A foundational principle for quantum mechanics. Found. Phys. **29**(4), 631–643 (1999)
12. J.S. Bell, On the Einstein Podolsky Rosen paradox. Physics Physique Fizika **1**(3), 195 (1964)
13. J.L. Cereceda, Local hidden-variable models and negative-probability measures (2020), arXiv: quant-ph/0010091 [quant-ph]
14. J.F. Clauser et al., Proposed experiment to test local hidden-variable theories. Phys. Rev. Lett. **23**(15), 880 (1969)
15. T. Maudlin, What Bell did. J. Phys. A Math. Theor. **47**(42), 424010 (2014)
16. A. Einstein, B. Podolsky, N. Rosen, Can quantum-mechanical description of physical reality be considered complete?. Phys. Rev. **47**(10), 777 (1935)
17. E. Schrödinger, Discussion of probability relations between separated systems, in *Mathematical Proceedings of the Cambridge Philosophical Society*, vol. 31. 4 (Cambridge University Press, 1935), pp. 555–563
18. N. Bohr, Can quantum-mechanical description of physical reality be considered complete?. Phys. Rev. **48**(8), 696 (1935)
19. L. Hardy, Quantum mechanics, local realistic theories, and Lorentz-invariant realistic theories. Phys. Rev. Lett. **68**(20), 2981 (1992)
20. L. Hardy, Nonlocality for two particles without inequalities for almost all entangled states. Phys. Rev. Lett. **71**(11), 1665 (1993)
21. H. Dourdent, *A Quantum Gödelian Hunch* (2020), arXiv: 2005.04274 [quant-ph]
22. D. Frauchiger, R. Renner, Quantum theory cannot consistently describe the use of itself. Nat. Commun. **9**(1), 1–10 (2018)
23. S. Aaronson, It's hard to think when someone Hadamards your brain. [Online; accessed 2020-10-21] (2018), https://www.scottaaronson.com/blog/?p=3975
24. E.P. Wigner, Remarks on the mind-body question, in *Philosophical Reflections and Syntheses* (Springer, 1995), pp. 247–260
25. J. Wheeler, Add "Participant" to "Undecidable Propositions" to arrive at Physics. [Online; accessed accessed 2020-10-21] (1974), https://jawarchive.files.wordpress.com/2012/03/twa-1974.pdf
26. J. Bernstein, *Quantum Proles* (Princeton University Press, Princeton, NJ, 1991)
27. A. Peres, Unperformed experiments have no results. Am. J. Phys. **46**(7), 745–747 (1978)

Chapter 9
Why Is the Universe Comprehensible?

Ian T. Durham

Abstract Why is the universe comprehensible? How is it that we can come to know its regularities well enough to exploit them for our own gain? In this essay I argue that the nature of our comprehension lies in the mutually agreed upon methodology we use to attain that comprehension and on the basic stability of the universe. But I also argue that the very act of comprehension itself places constraints on what we can comprehend by forcing us to establish a context for our knowledge. In this way the universe has managed to conspire to make itself objectively comprehensible to subjective observers.

9.1   Introduction

Why is the universe comprehensible? How is it that we can come to know its regularities well enough to exploit them for our own gain? This is one of the greatest enigmas of human knowledge. As Einstein once said, "[t]he eternally incomprehensible thing about the world is its comprehensibility" [9].[1] But, in order to comprehend something, Paul Davies has argued that the system being comprehended must necessarily have a certain level of organization that is embodied by the presence of non-random complexity within the system and that the act of comprehension requires accepting that this non-random complexity is in some sense objectively real [6]. Objective reality is necessarily observer-independent. It is an inherently third-person perspective. Yet, if something is said to be comprehensible it is worth asking by whom or what; comprehensibility by its very definition implies the existence of something or someone doing the comprehending. Thus it would seem that comprehension is an inherently first-person, subjective act. Somehow the universe has conspired to make itself objectively comprehensible to subjective observers.

---

[1] This translation of the phrase is attributed to Holton [18].

I. T. Durham (✉)
Department of Physics, Saint Anselm College, Manchester, NH 03102, USA
e-mail: idurham@anselm.edu

Yet, these two perspectives are not necessarily incompatible. For example, Markus Müller recently constructed a self-consistent theory in which an objective external world emerges from more fundamental observer states. The theory takes the first-person perspective as axiomatically true and, using the tools of algorithmic information theory, derives an emergent third-person perspective *from* it [20]. There is a sense in which a subjective first-person perspective is unavoidably fundamental. We are, after all, a part of this world that we seem to comprehend. We cannot step outside of it. In fact, our mere presence *within* it is surely a regularity worthy of an explanation. But that still does not explain how it is that we can even countenance the possibility of arriving at an explanation in the first place. The very fact that the concept of explanation exists is, itself, in need of some explanation.

To Einstein, 'comprehensibility' meant a scientific understanding of the universe's functional composition. That is, he is presumably not interested in any metaphysical considerations. While metaphysical arguments for comprehensibility may exist, cf. theological explanations, these offer little in the way of reproducibility. It is arguably reproducibility that bridges the gap between the first-person, subjective act of comprehension and the third-person objective reality that is being comprehended. In a certain sense it could be said that objective reality is *defined by* those regularities that can be independently reproduced or verified by different observers.

This question of the universe's comprehensibility, then, rests squarely on the ability of independent observers to reproduce or verify the existence of certain regularities. It is fundamentally an act of agreement, a willingness to acknowledge commonalities. In that sense, though comprehension is a subjective act for individual observers, for it to have any real meaning there must be some objective agreement by those observers about the regularities being comprehended. That is, the observers must, at least in a broad sense, share a common methodology for carrying out their observations. This makes the question of comprehensibility one of measurement. In this essay, I explore how the nature of this methodology itself helps to enable the comprehensibility of the universe while simultaneously placing limits on how much we can comprehend.

9.2 Comprehensibility

The famous philosophical question[2] "if a tree falls in a forest and no one is there to hear it fall, does it make a sound?" is usually interpreted as a metaphorical question about the nature of objective reality. Much less attention is paid to the nature of the question itself and what that implies for any hypothetical answer. A literal reading of the question and its sentence structure makes it clear that the question is about the sound produced (or not produced) by a falling tree. The tree is assumed to fall regardless of the presence or absence of a person in the vicinity. Even more

---

[2] The origins of this question appear to be in George Berkeley's *A Treatise Concerning the Principles of Human Knowledge* (1710). Its current form seems to have first been stated in [19].

fundamentally, there is never any question about the tree's existence. The tree exists and it has fallen. The question is whether the sound it makes in falling requires the presence of an observer.

Though the question concerning the tree is a truth-conditional sentence, not all questions necessarily take such a form. For example, the sentence "what color is my hair?"[3] does not have a truth value. However, the *answers* to questions are invariably declarative statements that *do* have truth values. Thus questions can nearly always be understood within truth-conditional semantics [5, 13]. But in order to judge the truth of a sentence we must understand its meaning [23]. According to Frege, there exist two primary types of expressions: proper names (which are nearly always singular terms) and functional expressions [11]. Understanding the meaning of a statement requires understanding the context of these expressions.

We can reformulate the question concerning the tree as a declarative statement to which a truth value may be assigned: "any tree that falls in a forest will make a sound regardless of the presence or absence of an observer." The proper name in this sentence is 'any tree that falls in a forest'. That is, this is a statement about a tree falling in a forest. It is *not* a statement about a dog or the color yellow. The functional expression places the falling tree in a forest in context by associating it with the predicate 'will make a sound regardless of the presence or absence of an observer'. Thus, this is a statement about whether a falling tree in a forest makes a sound in the presence or absence of an observer. It is *not* a statement about whether a falling tree in a forest turns into a pigeon or reads a book. The ability to assign a truth value to the full statement rests on the fact that these two expressions each have referents, i.e. they specify objects, conditions, etc. Though the referents are necessary in the formulation of the statement, they naturally constrain the truth value. Simply put, we could ask an infinite number of questions about a tree falling in a forest, but as soon as we choose which question to ask, we have narrowed the context. This may seem trivial but it is actually deeply profound. While the question "what color is my hair?" could garner an infinite number of responses (e.g. 'potato', 'narrow', 'blue', etc.) only a finite set of such responses makes logical sense. The statement "any tree that falls in a forest will believe in Santa Claus" is a nonsensical statement whose truth value has no real meaning.

In more concrete terms, consider an experiment designed to measure the spin of a single, free electron. One potential implementation of such an experiment consists of a source of single, free electrons (e.g. an electron gun), a Stern-Gerlach magnet, solid-state detectors, and a source of a transverse electric field to cancel out the transverse Lorentz force (see Fig. 9.1). This measurement asks and answers the question "what is the spin of an electron along the â axis?" We could restate the question as a truth-conditional statement of the form "the electron is spin-up along the â axis" (alternately, we could state it as spin-down along the same axis). Physically speaking, the 'proper name' (in Frege's terms) is the electron which is produced by the electron gun. All the other devices in the measurement generate the functional expression that places the electron in the context of a spin measurement along the â

---

[3] This is a trick question. I'm bald.

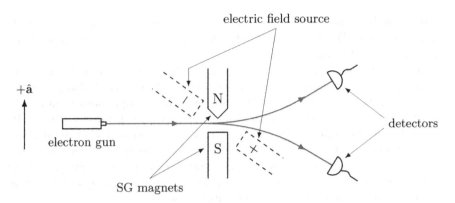

electric field source

+â

electron gun

N

S

SG magnets

detectors

Fig. 9.1 One method of measuring the spin of single electrons along a particular axis **a**, involves firing a properly attenuated beam of electrons through a Stern-Gerlach (SG) magnet and an appropriately applied transverse electric field. The transverse electric field is required to balance the Lorentz force. Each electron is then incident on one of two solid-state detectors

axis. That is, the choice of devices inherently determines the subject and predicate of the experiment. This experiment as it has just been described, for example, will not measure the spin of an electron along some axis $\hat{b}$. In order to measure spin along an axis $\hat{b}$, the magnetic and electric fields would need to be appropriately rotated. This can only be accomplished by modifying the apparatus which necessarily changes the functional expression. Likewise, neither will this experiment measure the spin of a silver atom along *any* axis. This is not because the devices representing the functional expression can't accomplish the task, but rather it is because the source representing the proper noun (in Frege's terminology) only produces electrons. Even more obviously, this experiment will never determine the color of my hair without considerable modification.

Of course, one could simply deny these limitations and declare that the experiment *has* actually determined the spin of an electron along some axis $\hat{b}$ or the spin of a silver atom or the color of my hair. That is, an individual observer could have an entirely different semantic interpretation of the experiment. This gets at the heart of science itself. As an anonymous reviewer in *Philosophical Magazine* once declared (as quoted in [8]), "[s]cience is the rational correlation of experience." That is, the *objective* power of science lies in the correlation of multiple *subjective* experiences. Each individual observer is free to declare whatever he or she wishes to declare. It is only when observers come to agree on elements of experience through rational means that objectivity is gained.[4] Consider, for example, an experiment designed to test a Bell-type inequality using electron spin. We begin with a source of entangled electron pairs where the pairs are emitted in the form of two electron beams. Each beam passes through a setup similar to the one shown in Fig. 9.1 except the observer operating the equipment on a particular beam has the freedom to rotate the magnetic

---

[4] Due to length restrictions I will not endeavor to define 'rational means' here. In relation to the aforementioned quote, Eddington expands on the idea in [8].

and electric fields to any desired direction thereby allowing for a measurement of spin along any axis in a plane orthogonal to the beam. By convention we refer to the two observers, each associated with one of the two beams, as Alice and Bob. Crucially, in order for this experiment to produce meaningful results, Alice and Bob must agree on the measurement protocol. That is, if their goal is to test a Bell-type inequality in such a way that they both agree on the result, then they must agree to the semantics of the statement whose truth value the experiment aims to determine. For instance, if Alice chooses to measure the spin of her electron along an axis $\hat{\mathbf{b}}$, she and Bob must have agreed to the definition of axis $\hat{\mathbf{b}}$ prior to carrying out the experiment. If they have not agreed to the definition of axis $\hat{\mathbf{b}}$ beforehand, they have little hope of finding a meaningful result to the experiment. But once they *do* agree to the definition of axis $\hat{\mathbf{b}}$, they will have necessarily limited the context of the experiment since they will have defined axis $\hat{\mathbf{b}}$ to correspond to a specific direction that they both agree on. If they wish to achieve their stated goal of testing the Bell-type inequality, neither of them can simply deny the limitations of the setup. They *must* agree on the semantic interpretation of the proper noun and functional expression in any truth-conditional statement they wish to jointly test *since that is the point of the experiment*. This is another deeply profound point.

In a single-observer experiment, the observer is free to assign any meaning at all to the semantic content of the experimental statement. In the experiment described in Fig. 9.1, for example, the observer could declare that the source produces silver atoms or even trees, for that matter, and simply assume that the results make meaningful sense. In other words, to a single observer, the truth value of the truth-conditional statement that the observer is testing does *not* necessarily depend on that observer's interpretation of the statement. Observers in isolation are free to simply deny any result they obtain or interpret it in any way they wish. Absent any independent verification, they could simply declare that the experiment produced a particular truth value. But in a two-observer experiment such as a Bell-type inequality test, the truth value of the statement being tested *necessarily* depends on the mutually agreed upon interpretation of the statement they are testing. Suppose, for example, that Alice and Bob *do* agree on the interpretation of the statement they are testing and suppose that statement is something to the effect of "a series of N runs of this experiment violates a Bell inequality." This means that they will agree on what constitutes a single run of the experiment, which Bell inequality is being tested (what a Bell inequality even *is*, for that matter), what constitutes axis $\hat{\mathbf{a}}$, what constitutes axis $\hat{\mathbf{b}}$, etc. If, under these conditions, they act in good faith, then it means that they will agree on the truth value of the statement that they are testing. Note that if they have agreed to all these points (which constitute the functional expression of the statement) then (acting in good faith) they will necessarily agree on the result *even if the equipment fails*. This is because they have agreed to the form of the functional expression itself. If you and I agree that a certain switch operates a light and we agree on what it means for that light to be illuminated when the switch is in a certain state, then we will agree (if acting in good faith) if the light *fails* to illuminate when the switch is in this state. We may disagree as to *why* it failed to illuminate, but by agreeing to the conditions of the experiment we should agree on whether or not it *has* illuminated. If we do not, then either one of us is not acting in good faith or there is some potentially hidden source of disagreement in the setup or execution.

Put another way, on a subsequent run of the experiment, Alice and Bob could fail to reach an agreement on what constitutes axis $\hat{\mathbf{b}}$. This disagreement could be known to them, but it also could be the result of experimental error. In either case, it is conceivable that Alice could find a violation of the inequality while Bob does not. This leaves the truth value of the statement that they are testing ambiguous. Each can *say* that they individually comprehend the results of the experiment, but without some form of agreement, their individual interpretations remain nothing more than metaphysical speculation. That is, for their individual subjective comprehension to have any objective, scientific power, they must agree to what is being comprehended which means they must agree on the interpretation of the statement that they are testing. This is the only way in which the experiment can produce an unambiguous truth value. In other words, while individual observers can claim to observe anything, in order for truth values of statements to be unambiguous, observers must agree to the context of these statements and to the protocol that is to be used to determine those truth values. In fact, this is really the heart of reproducibility in science. If an experiment cannot be reproduced, there must be a disagreement, either in the interpretation of the result or in the setup or operation of the experiment.

The comprehension that Einstein speaks of, then, arises from two interdependent conditions: (1) observers (of which there must be more than one) must agree on the meaning of truth-conditional statements and the methods used to test those statements, and (2) the results of the agreed upon tests must, in the aggregate, remain within the physical context of the tests themselves, i.e. an experiment such as the one depicted in Fig. 9.1, for example, will not determine the color of my hair. In a certain sense, assuming a stable universe, the second condition follows from the first. If Alice and Bob agree to the meaning of a question, they immediately limit the number of possible responses that will make any logical sense. If they then agree to the method of testing the question, they further limit the possible answers to be within the context of the test methodology. Put simply, if Alice and Bob agree to ask the question "what color is my hair?" and they agree on the methodology for testing that question, they can be fairly confident that the answer they find will not be 'narrow' or 'dog'. As long as the universe remains relatively stable, the first condition implies the second.

The fact that the universe remains relatively stable, thus ensuring that the second condition is met in the aggregate, was originally proposed as a "principle" of comprehensibility in [7]. It asserted that the nature of a physical system under investigation will always remain within the bounds of the method of investigation. That is, we expect scientific answers to scientific questions. Of course, one could object to the use of the word 'always' as there is no way to prove this. In fact it is really a statement of tendencies within physical systems akin to the Second Law of Thermodynamics. In terms of truth-conditional statements, then, we can restate this principle of comprehensibility as follows.

Principle of Comprehensibility (Truth-conditional form) The results produced by tests of truth-conditional statements, for which the statements and the method of testing those statements are both agreed-upon by two or more observers, tend to remain within the physical context of the tests themselves.

Again, despite being written here in terms of truth-conditional statements, which have a formal structure, this is a *physical* principle. This is why we specify a '*physical* context' within the principle's statement. It is not a formal (logico-mathematical) statement and thus has no rigorous proof. While it is impossible to *prove* that the answer to the question "what color is my hair?" will never be 'narrow' or 'dog', the fact remains that it is highly unlikely. Even more unlikely is that 'narrow' or 'dog' will be an outcome of the experiment described in Fig. 9.1. In short, stability and context lead to comprehension. Comprehension, however, comes with a price.

9.3   The Price of Comprehensibility

By simply asking a question, we immediately establish a context which limits the scope of the inquiry. The general stability of the universe ensures that potential answers to our question will, to a high degree of probability, remain within the context of the question. This is the essence of comprehensibility since it allows us to develop systematic measures by which we can further probe a topic. Building an understanding requires refining the context by asking additional questions. But this also necessarily means that our knowledge is shaped by the questions we ask and how we choose to ask them, i.e. by our methodology.

As a simple example, suppose we are tasked with designing a device that will compute a number by solving a specific equation. Choosing a design for the fundamental operation of our device is analogous to choosing a set of truth-conditional statements for a particular question. So, for example, perhaps we designed our device to be decimal (i.e. base-ten) at its most fundamental and it computes the number 1197. How this number is represented to us is immaterial. What matters is that the device was able to compute the number *exactly* based on how it was designed. On the other hand, we could have designed our device to be binary at its most fundamental, in which case it would compute the number 10010101101 since $10010101101_2 = 1197_{10}$. Again, the representation of the number to us is immaterial here. Both devices could represent the number to us as base-eight or base-six or even as sounds or images. What matters is that the two designs are fundamentally different yet both ultimately produce the same solution to the equation with equal accuracy. The principle of comprehensibility ensures that the decimal device performs decimal operations while the binary device performs binary operations, i.e. the devices produce what they are *designed* to produce. If they reliably and consistently continue to do so, we might infer that this pattern represents some kind of objectively real aspect of the world and we could say that we comprehended some element of it.

But suppose, instead, that our decimal device finds that the solution to the input equation is 1/10 and let us further suppose that this answer is exact. If our device was instead binary, the best it could do would be to approximate this number as a non-terminating expansion,

$$\frac{1}{10} \approx \frac{1}{16} + \frac{1}{32} + \frac{1}{256} + \cdots = \frac{1}{2^4} + \frac{1}{2^5} + \frac{1}{2^8} + \cdots$$

That is, the design of our device can potentially limit the accuracy of our knowledge and constrain our comprehension. If we are in doubt about the results we obtain, we could theoretically design a different device and check the results. If we began with our binary device and realized the result was an approximation, we might be able to design a decimal device and compare the results of the two. If the decimal device naturally gives a fixed, terminating answer we can assume it is more accurate. But if we are, say, computing π, how would we know which is more accurate?

This is not a trivial point. John Wrench and Levi Smith famously calculated the digits of π using a gear-driven calculator by solving Machin's formula, eventually reaching 1120 digits by 1949 [22]. Since then all improvements to their estimation have been on electronic computers. While decimal computers do exist and have been used to search for the digits of π (e.g. the ENIAC), most of the machines used to calculate π have been binary in their basic functionality and simply encoded decimal digits using some scheme such as binary-coded decimal (BCD) or excess-3. Further, since 2009 all records for the digits of π have been carried out using Alexander Yee's multi-threaded y-cruncher program.[5] So not only have we limited ourselves to electronic computers in our search for the digits of π but we have recently limited ourselves to a single *algorithm*. But as the simple example above demonstrates, the choice of *device itself* can constrain the accuracy of the results. How do we know that a mechanical computer would necessarily give the same set of digits? After all, a mechanical computer would not likely be able to execute the algorithm that is at the heart of Yee's y-cruncher program since that program is designed expressly for digital computers. But constructing a mechanical computer to calculate 50 trillion digits of π (the current record) would likely be extremely difficult if not impossible. There are other reasons to trust Yee's algorithm and electronic computations in general, but the point is that there is really no way to *prove* just how accurate they are in an experimental sense. Simply put, in order to compute a number (like π) we must build a device capable of carrying out the computation. But in doing so, we must choose a design and that choice immediately sets constraints on what types of results we can expect to get from the device. A binary device will produce binary results while a decimal device will produce decimal results.

The Church-Turing thesis conjectures that it ultimately should not matter what form the computation takes. The types of problems computable using one model should be computable using *any* model since the thesis asserts that all computational models are equivalent to Turing machines. But the Church-Turing thesis is a formal

---

[5] For details see http://www.numberworld.org/y-cruncher/.

conjecture. It's unclear if such a conjecture can necessarily be applied to physical systems [3, 4] though several proposals for physical versions of the thesis have been made [1, 21, 27]. But the physical world is a fickle beast. Proving something physically is very different from proving something mathematically. This point goes well beyond merely addressing Hilbert's Sixth Problem in which he called for an extension of the axiomatic methods of mathematics to physics [15]. It is conceivable that, should a suitable theory of quantum gravity be found, we could eventually see a fully axiomatized physical *theory*. But anyone who has spent any time in a laboratory will attest to the fact that the real world is far messier than theory would have us believe.

Comprehensibility is about what we, as humans, can know for certain about the universe and it is through experiment that we do this. As Richard Feynman once famously said, if something disagrees with experiment then it's wrong [10]. But the nature of comprehensibility ensures that the answers to the questions we pose to the universe are constrained by the form of the question itself. A *physical* binary device, i.e. not an idealized abstract one, will always produce binary results and thus, for example, will never be able to calculate $1/10$ exactly. Certain *physical* computations will always be approximate.

This has an even deeper and more profound implication if science ultimately rests entirely on proofs of truth-conditional statements. Proving a truth-conditional statement is equivalent to solving a decision problem or what Hilbert and Wilhelm Ackermann referred to as the *Entscheidungsproblem* [16]. Such a problem produces a yes-or-no answer to a question given any number of inputs. John Wheeler famously argued that the answers to such questions were the basis of all that exists [26]. As late as 1930 Hilbert believed there would never be an unsolvable problem [17]. Yet, though the Church-Turing thesis conjectures that all computational models are equivalent to Turing machines, it also allows for the existence of noncomputable functions. That is, it allows for the existence of decision problems for which no algorithm can be constructed that is guaranteed to lead to a correct yes-or-no answer. Such problems are said to be undecidable. Turing, for instance, famously showed that the halting problem was undecidable on Turing machines [24, 25]. But all of this work was grounded in formal methods. If the Principle of Comprehensibility is valid it would seem to imply that there might exist problems that are undecidable for *physical* reasons. That is, if Wheeler is right and all that exists derives its very existence from "apparatus-elicited" answers to yes-or-no questions, then there are elements of the physical universe that are simply unknowable. Furthermore, the origin of that unknowability is both logical *and* physical. Some aspects might be unknowable because we cannot construct an algorithm that is guaranteed to lead to a correct truth value for some truth-conditional statements. Other aspects might be unknowable because the universe's fundamental fabric is such that no machine can be constructed to produce a correct truth value for some truth-conditional statements. These are distinct points unless the universe itself is a Turing machine. The latter follows from the constraints associated with comprehensibility. In order to answer a question, we must choose a set of truth-conditional statements that we will test and agree to a testing methodology. This ensures that the results will be comprehensible.

But, by choosing a particular set of such statements as opposed to some other set, we are shaping the form in which our answer will appear. When we ask a question we are necessarily carving out a small portion of knowledge from all that it is possible to know. Yet the question we ask and how we ask it immediately contextualizes any potential answer (even one that doesn't logically follow) and thus shapes any knowledge we gain from it. There is simply no way around this.

9.4   Limitations

The above arguments largely rest on the ability to reformulate the questions that lie at the heart of the scientific enterprise in terms of truth-conditional statements. But can all questions be reduced to truth-conditional statements? Can the scientific enterprise even be fully reduced to a set of questions? Attempts to formulate scientific explanation in terms of the answers to questions has a long history. Hempel and Oppenheim argued that all scientific explanations could be regarded as answers to 'why-questions' such as "[w]hy do the planets move in elliptical orbits with the Sun at one focus?" [14]. Often referred to as the *deductive-nomological* (DN) model, their account of the scientific enterprise is one of deduction that is reliant on the accurate prediction or postdiction of phenomena. But this model fell out of favor in the philosophy of science community for several decades and, despite a resurgence, continues to have its detractors chiefly because it appears to exclude some types of explanations generally regarded as scientific [12]. But even assuming that the scientific enterprise can be reduced to a set of questions, it is not clear that it can be further reduced to sets of truth-conditional statements. David Braun, for instance, offers an alternative in which the answers to questions simply provide contextual information [2]. Truth is a loaded term. After all, how do we know that our knowledge of the world is even real? Would it not be better to simply speak of information? In Newton's day, testing the truth-conditional statement "time is absolute" would have produced a positive truth value given the knowledge and technology of the time. That same statement, if tested now, would elicit a negative truth value. Yet it is wrong to say that there was any change in the underlying physics between then and now. What changed was our *knowledge* of that physics, i.e. we increased our information.

Yet even if we take Braun's view it is clear that limitations still exist and that these limitations intimately depend on context. Our inability to simultaneously measure non-commuting observables in quantum systems to arbitrary accuracy is a limit on our ability to obtain information regardless of whether or not we believe in objective truth. It arises from the context of our measurement. Comprehensibility is still the result of a combination of the mutual agreement between observers and the fact that the universe remains relatively stable. That is to say, the Principle of Comprehensibility need not be formulated in a truth-conditional form. The fact remains that in order to say we comprehend some element of the universe we must necessarily obtain some information about that element. But obtaining that information is a physical process that necessarily has a context which constrains the nature of that information; the very act of acquiring information shapes the information acquired. Physical limitations on the acquisition of knowledge are not controversial. The universe has fundamental limits baked into it. But it is these very limits that allow for the universe to be comprehensible. They are necessary in order for our seemingly finite minds to have any hope of comprehending anything.

These physical limitations bear a certain resemblance to Gödel's incompleteness theorems in that they arise from the internal structure of the system and one must break free from that structure in order to fully understand it. The universe is a vast and interconnected place of which we are but a small part, a mere mote of dust, as it were. Any attempt to comprehend it must necessarily depend on the fact that we are a part of it. Indeed the very act of comprehension is *itself* a part of it and is thus shaped by it. Like the god Odin from Norse mythology, who is said to have sacrificed an eye in order to attain wisdom, our quest for comprehension limits our very ability to comprehend, and the universe remains always partially veiled.

References

1. P. Arrighi, G. Dowek, The physical Church-Turing thesis and the principles of quantum theory. Int. J. Found. Comput. Sci. **23**(5), 1131–1145 (2012)
2. D. Braun, Now you know who Hong Oak Yun is. Philos. Issues **16**, 24–42 (2006)
3. B.J. Copeland, Computation, in *The Blackwell Guide to the Philosophy of Computing and Information*, ed. by L. Floridi (Wiley-Blackwell, Hoboken, 2004)
4. B.J. Copeland, O. Shagrir, The Church-Turing thesis: logical limit or breachable barrier? Commun. ACM **62**(1), 66–74 (2019)
5. C. Cross, F. Roelofsen, Questions, in *The Stanford Encyclopedia of Philosophy*, ed. by E.N. Zalta (Metaphysics Research Lab, Stanford University, Spring, 2018 edition, 2018)
6. P.C.W. Davies, Why is the Physical World so comprehensible? in *Complexity*, Entropy and the Physics of Information, Santa Fe Institute Studies in the Science of Complexity, ed. by W.H. Zurek (Addison-Wesley, Redwood City, CA, 1990), pp. 61–70
7. I.T. Durham, Boundaries of scientific thought, in *Information and Interaction: Eddington, Wheeler, and the Limits of Knowledge*, The Frontiers Collection, ed. by I.T. Durham, D. Rickles (Springer, Cham, 2017), pp. 1–34
8. A.S. Eddington, *The Philosophy of Physical Science* (Cambridge University Press, Cambridge, 1939)
9. A. Einstein, Physik und Realität. J. Frankl. Inst. **221**, 313–382 (1936)
10. R.P. Feynman, *The Character of Physical Law* (Modern Library, New York, 1965)
11. G. Frege, Function and concept, in *Translations from the Philosophical Writings of Gottlob Frege*, Oxford Readings in Philosophy, ed. by P. Geach, M. Black, 3rd edn. (Basil Blackwell, Oxford, 1980), pp. 130–149
12. S. Glennan, Explanation, in *Philosophy of Science*, ed. by S. Sarkar, J. Pfeiffer (Routledge, Abingdon, 2006)
13. C.L. Hamblin, Questions. Australas. J. Philos. **36**, 159–168 (1958)
14. C. Hempel, P. Oppenheim, Studies in the logic of explanation. Philos. Sci. **15**(2), 135–175 (1948)
15. D. Hilbert, Mathematical problems. Bull. Am. Math. Soc. **8**(10), 437–479 (1902)
16. D. Hilbert, W. Ackermann, *Grundzüge der theoretischen Llogik* (Springer, Berlin, 1928)
17. A. Hodges, *Alan Turing: The Enigma* (Simon and Schuster, New York, 1983)
18. G. Holton, What precisely is "thinking"? ...Einstein's answer. Phys. Teach. **17**, 157 (1979)

19. C.R. Mann, G.R. Twiss, *Physics* (Foresman and Co., Chicago, 1910)
20. M. Müller, Law without law: from observer states to physics via algorithmic information theory (2019), https://arxiv.org/abs/1712.01826
21. G. Piccinini, The physical Church-Turing thesis: modest or bold? Br. J. Philos. Sci. **62**(4), 733–769 (2011)
22. D.S. Richeson, *Tales of Impossibility: The 2000-Year Quest to Solve the Mathematical Problems of Antiquity* (Princeton University Press, Princeton and Oxford, 2019)
23. A. Tarski, The semantic conception of truth and the foundations of semantics. Philos. Phenomenol. Res. **4** (1944)
24. A.M. Turing, On computable numbers, with an application to the Entscheidungsproblem. Proc. Lond. Math. Soc. **42**(2), 230–265 (1937)
25. A.M. Turing, On computable numbers, with an application to the Entscheidungsproblem. A Correction. Proc. Lond. Math. Soc. **43**(2), 544–546 (1938)
26. J. Wheeler, Information, physics, quantum: the search for links, in *Complexity, Entropy and the Physics of Information*, ed. by W.H. Zurek (Addison-Wesley, Redwood City, CA, 1990), pp. 3–28
27. S. Wolfram, *A New Kind of Science* (Wolfram Media, Champaign, 2002)

Chapter 10
Noisy Deductive Reasoning: How Humans Construct Math, and How Math Constructs Universes

David H. Wolpert and David Kinney

Abstract We present a computational model of mathematical reasoning according to which mathematics is a fundamentally stochastic process. That is, in our model, whether or not a given formula is deemed a theorem in some axiomatic system is not a matter of certainty, but is instead governed by a probability distribution. We then show that this framework gives a compelling account of several aspects of mathematical practice. These include: 1) the way in which mathematicians generate research programs, 2) the applicability of Bayesian models of mathematical heuristics, 3) the role of abductive reasoning in mathematics, 4) the way in which multiple proofs of a proposition can strengthen our degree of belief in that proposition, and 5) the nature of the hypothesis that there are multiple formal systems that are isomorphic to physically possible universes. Thus, by embracing a model of mathematics as not perfectly predictable, we generate a new and fruitful perspective on the epistemology and practice of mathematics.

10.1 Introduction

Humans are imperfect reasoners. In particular, humans are imperfect *mathematical* reasoners. They are fallible, with a non-zero probability of making a mistake in

D. H. Wolpert (✉) · D. Kinney (✉)
Santa Fe Institute, Santa Fe, NM, USA
e-mail: david.h.wolpert@gmail.com
URL: http://davidwolpert.weebly.com

D. Kinney
URL: http://davidbkinney.com

D. H. Wolpert
Complexity Science Hub Vienna, Vienna, Austria

Arizona State University, Tempe, AZ, USA

any step of their reasoning. This means that there is a nonzero probability that any conclusion that they come to is mistaken. This is true no matter how convinced they are of that conclusion. Even brilliant mathematicians behave in this way; Poincaré wrote that he was "absolutely incapable of adding without mistakes" ([14], p. 323).

The mirthful banter of Poincaré aside, such unavoidable noise in human mathematical reasoning has some far-reaching consequences. An argument that goes back (at least) to Hume points out that since individual mathematicians are imperfect reasoners, the entire community of working mathematicians must also be one big, imperfect reasoner. This implies that there must be nonzero probability of a mistake in every conclusion that mathematicians have ever reached (Hume [10], Viteri and DeDeo [21]). This noise in the output of communal mathematical research is *unavoidable*, inherent to any physical system (like a collection of human brains) that engages in mathematical reasoning. Indeed, one might argue that there will also be unavoidable noise in the mathematics constructed by any far-future, post-singularity hive of AI mathematicians, or by any society of demi-God aliens whose civilization is a billion years old. After all, awe-inspiring as those minds might be, they are still physical systems, subject to nonzero noise in the physical processes that underlie their reasoning.

By contrast, almost all work on the foundations and philosophy of mathematics to date has presumed that mathematics is the product of noise-*free* deductive reasoning. As Hilbert [9] famously said, "mathematical existence is merely freedom from contradiction".

In light of this discrepancy between the actual nature of mathematics constructed by physically-embodied intelligences and the traditional view of mathematics as noise-free, here we consider the consequences if we abandon the traditional view of "mathematical existence" as noise-free. We make a small leap, and identify what might be produced by any community of far-future, galaxy-spanning mathematicians as *mathematics itself*. We ask, what are the implications if mathematics itself, abstracted from any particular set of physical reasoners, is a stochastic system? What are the implications if we represent mathematics not only as inescapably subject to instances of undecidability and uncomputability, as Gödel [7] first showed, but also inescapably *unpredictable* in its conclusions, since it is actually stochastic?

In fact, if you just ask them, many practicing human mathematicians *will tell you* that there is a broad probability distribution over mathematical truths. For example, if you ask them about any Clay prize question, most practicing mathematicians would say that any of the possible answers has nonzero probability of being correct. What if mathematicians are right to say there is a broad distribution over mathematical truths, not simply as a statement about their subjective uncertainty, but as a statement about mathematical reality? What if there is a non-degenerate *objective* probability distribution over mathematical truths, a distribution which "is the way things really are", independent of human uncertainty? What if in this regard mathematics is just like quantum physics, in which there are objective probability distributions, distributions which are "the way things really are", independent of human uncertainty?

In this essay we present a model of mathematical reasoning as a fundamentally stochastic process, and therefore of mathematics itself as a fundamentally stochastic system. We also present a (very) preliminary investigation of some of this model's features. In particular, we show that this model:

- allows us to formalize the process by which actual mathematical researchers select questions to investigate.
- provides a Bayesian justification for the role that abductive reasoning plays in actual mathematical research.
- provides a Bayesian justification of the idea that a mathematical claim warrants a higher degree of belief if there are multiple lines of reasoning supporting that claim.
- can be used to investigate the mathematical multiverse hypothesis (i.e., the hypothesis that there are multiple physical realities, each of which is isomorphic to a formal system) thereby integrating the analysis of the inherent uncertainty in the laws of physics with analysis of the inherent uncertainty in the laws of mathematics.

If mathematics is "invented" by human mathematicians, then it obviously *is* a stochastic system, and should be modeled as such. (In this case, the distributions of mathematics are set by the inherent noise in human mathematical reasoning.) Going beyond this, we argue that even if mathematics is "discovered" rather than invented, it may still prove fruitful to weaken the a priori assumption that what is being discovered is noise-free—just as it has often proven fruitful in the past to weaken other assumptions imposed upon mathematics. In this essay, we start to explore the implications if mathematics is a stochastic system, without advocating either that it is invented or that it is discovered—as described below, our investigation has implications in both cases.[1]

10.2 Formal Systems

The concept of a "mathematical system" can be defined in several equivalent ways, e.g., in terms of model theory, Turing machines, formal systems, etc. Here we will follow Tegmark [17] and use formal systems. Specifically, a (**recursive**) **formal system** can be summarized as any triple of the form

1. A finite collection of symbols, (called an **alphabet**), which can be concatenated into **strings**.
2. A (recursive) set of rules for determining which strings are **well-formed formulas** (WFFs).
3. A (recursive) set of rules for determining which WFFs are **theorems**.

---

[1] Note that just like the authors of all other papers written about mathematics, we believe that the deductive reasoning in this essay is correct. The fact that we acknowledge the possibility of erroneous deductive reasoning, and that in fact the unavoidability of erroneous reasoning is the topic of this essay, doesn't render our belief in the correctness of our reasoning about that topic any more or less legitimate than the analogous belief by those other authors.

As considered in Tegmark [17, 18], formal systems are equivalence classes, defined by all possible automorphisms of the symbols in the alphabet. A related point is that strictly speaking, if we change the alphabet then we change the formal system. To circumvent such issues, here we just assume that there is some large set of symbols that contains the alphabets of all formal systems of interest, and define our formal systems in terms of that alphabet. Similarly, for current purposes, it would take us too far afield to rigorously formalize what we mean by the term "rule" in (2, 3). In particular, here we take rules to include both what are called "inference rules" and "axioms" in Tegmark [17].

As an example, standard arithmetic can be represented as a formal system [17]. '$1 + 1 = 2$' is a concatenation of five symbols from the associated alphabet into a string. In the conventional formal system representing standard arithmetic, '$1 + 1 = 2$' is both a WFF and a theorem. However, '$+4-$' is not a WFF in that formal system, despite being a string of symbols from its alphabet.

The community of real-world mathematicians does not spend their days just generating theorems in various formal systems. Rather they pose "open questions" in various formal systems, which they try to "answer". To model this, here we restrict attention to formal systems that contain the Boolean $\sim$ (NOT) symbol, with its usual meaning. If in a given such formal system a particular WFF φ is not a theorem, but $\sim \varphi$ is a theorem, we say that φ is an **antitheorem**. For example, '$1 + 1 = 3$' is an antitheorem in standard arithmetic. Loosely speaking, we formalize the "open questions" of current mathematics as pairs of a formal system S together with a WFF in S, φ, where mathematicians would like to conclude that φ is either a theorem or an antitheorem. Sometimes, φ will be a WFF in S but neither a theorem nor an antitheorem. We call such strings φ **undecidable**. As an example, Gödel [7] showed that any formal system strong enough to axiomatize arithmetic must contain undecidable WFFs.

To use these definitions to capture the focus of mathematicians on "open questions", in this essay we re-express formal systems as pairs rather than triples:

1. An alphabet;
2. A recursive set of rules for assigning one of four **valences** to all possible strings of symbols in that alphabet: 'theorem (t)', 'antitheorem (a)', 'not a WFF (n)', or 'undecidable (u)'.

It will be convenient to refer to any pair (S, φ) where S is a formal system and φ is a string in the alphabet of S as a **question**, and write it generically as q. We will also refer to any pair (q, v) where v is a valence as a **claim**.

10.3   A Stochastic Mathematical Reasoner

The *physical Church-Turing thesis* (PCT) states that the set of functions computable by Turing machines (TMs) include all those functions "that are computable using mechanical algorithmic procedures admissible by the laws of physics" ([24], p. 17).

If we assume that any mathematician's brain is bound by the laws of physics, and so their reasoning is also so bound, it follows that any reasoning by a mathematician may be emulated by a TM. However, as discussed above, we wish to allow the reasoning of human mathematicians to be inherently stochastic. In addition, since a TM is itself a system for carrying out mathematical reasoning, we want to allow the operation of a TM to be stochastic.

Accordingly, in this essay we amend the PCT to suppose that any reasoning by mathematical reasoner—human or otherwise—may be emulated by a special type of *probabilistic* Turing Machine (PTM) (see appendix for discussion of TMs and PTMs). We refer to PTMs of this special type as **noisy deductive reasoning machines** (NDR machines). Any NDR machine has several tapes. The **questions tape** always contains a finite sequence of unambiguously delineated questions (specified using any convenient, implicit code over bit strings). We write such a sequence as Q, and interpret it as the set of all "open questions currently being considered by the community of mathematicians" at any iteration of the NDR machine. The separate **claims tape** always contains a finite sequence of unambiguously delineated claims, which we refer to as a **claims list**. We write the claims list as C, and interpret it as the set of all claims "currently accepted by the community of mathematicians" at any iteration of the NDR machine. In addition to the questions and claims tapes, any NDR machine that models the community of real human mathematicians in any detail will have many work tapes, but we do not need to consider such tapes here (Fig. 10.1).

The NDR machine starts with the questions and claims tapes blank. Then the NDR machine iterates a sequence of three steps. In the first step, it adds new questions to Q. In the second step the NDR machine "tries" to determine the valences of the questions in Q. In the third step, if the valence v of one or more questions q has been found, then the pair (q, v) is added to the end of C, and q is removed from Q. We also allow the possibility that some claims in C are removed in this third step. The NDR machine iterates this sequence of three steps forever, i.e., it never halts. In this way the NDR machine randomly produces sequences of claims lists. We write the (random) claims list produced by an NDR machine after k iterations as C^k, generated by a distribution P^k. (Note that $P^k(C)$ can be nonzero even if $|C| \neq k$, i.e., if the number of claims in C differs from k.)

As an illustration, for any NDR machine that accurately models the real community of practicing mathematicians, the precise sequence of questions in the current claims list C must have been generated in a somewhat random manner, reflecting randomness in which questions the community of mathematicians happened to consider first. The NDR machine models that randomness in the update distribution of the underlying PTM. In addition, in that NDR machine it is extremely improbable that a claim on the claims tape ever gets removed.

There are several restrictions on NDR machines which are natural to impose in certain circumstances, especially when using NDR machines to model the community of human mathematicians. In particular, we say that a claims list C is **non-repeating** if it does not contain two claims that have the same question, otherwise it is **repeating**. We say that an NDR machine is non-repeating if it produces non-repeating claims

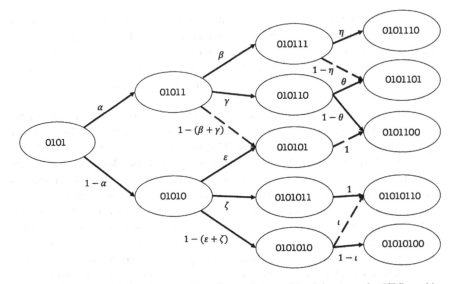

Fig. 10.1 Directed graph showing several possible evolutions of the claims tape of an NDR machine for a binary alphabet. Dashed arrows denote both the deletion of bits on the claims tape and concatenation of additional bits onto the claims tape, whereas solid arrows denote only the concatenation of bits. Labels on arrows show transition probabilities from each claims list to the next, which are determined by the update distribution of the NDR machine

lists with probability 1. As an example, if the NDR machine of the community of mathematicians is non-repeating, then there might be hidden contradictions lurking in the set of all claims currently accepted by mathematicians, but there are not any *explicit* contradictions.

For each counting number n, let C_n be the set of all sequences of n claims. For any current C and any $n \leq |C|$, define $C(n)$ to be the sequence of the first n claims in C. We say that a finite claims list C is **mistake-free** if for every claim $(q, v) \in C$, v is either t, a, n, u, depending on whether the question q is t, a, n or u, respectively. In other words, a claims list is mistake-free if for every claim (q, v) in that list, if $q = (\mathcal{S}, \varphi)$, then v is the syntactic valence assigned to φ by $\mathcal{S}$. As an example, most (all?) current mathematicians view the "currently accepted body of mathematics" as a mistake-free claims list. (However, even if it so happened that that current claims list actually were mistake-free, we do *not* assume that humans can determine that fact; in fact, we presume that humans cannot make that determination in many instances.) We say that an NDR machine is mistake-free if for all finite n, the probability is 1 that any claims list C produced by the NDR machine will be mistake-free.

We want to analyze the stochastic properties of the claims list, in the limit that the mathematical reasoner has been running for very many iterations. To do that, we require that for any n, the probability distribution of sequences of claims $C^k(n) \in C_n$ at the beginning of the claims list C that has been produced by the NDR machine at its k'th iteration after starting from its initial state converges in probability in the limit of $k \to \infty$. We also require that the set of all repeating claims lists has probability 0 under that limiting distribution. (Note though that we do not forbid repeating claims lists for finite k.) We further require that for all $n > 0$, the infinite k limit of the distribution over $C^k(n)$ is given by marginalizing the last (most recent) claim in the infinite k limit of the distribution over $C^k(n+1)$.[2] We write those limiting distributions as $P^\infty(C(n))$, one such distribution for each n.

For each n, the distribution $P^\infty(C(n))$ over all n-element claim sequences defines a probability distribution over all (unordered) **claims sets** $c = \{c_i\}$ containing $m \leq n$ claims:

$$P_n^\infty(c) := \sum_{C(n):\forall i, c_i \in C(n)} P^\infty(C(n)) \tag{10.1}$$

(where $c_i \in C(n)$ means that claim c_i occurs as one of the claims in the sequence $C(n)$). Under the assumptions of this essay, the $n \to \infty$ limit of this distribution over claims sets of size $m \leq n$ specifies an associated distribution over all finite claims sets, i.e., $\lim_{n \to \infty} P_n^\infty(c)$ is well-defined for any fixed, finite claims set c. We refer to this limiting distribution as the **claims distribution** of the underlying NDR machine, and write it as $\overline{P}(c)$. Intuitively, the claims distribution is the probability distribution over all possible bodies of mathematics that could end up being produced if current mathematicians kept working forever.[3] We say that a claims list (resp., claims set) is **maximal** if it has nonzero probability under P^∞ (resp., $\overline{P}$), and if it is not properly contained in a larger claims list (resp., claims set) that has nonzero probability.

Due to our assumption that there is zero probability of a repeating claims list under the claims distribution, the conditional distribution

$$\overline{P}(v \mid q) := \frac{\overline{P}((q, v))}{\overline{P}(q)} \tag{10.2}$$

$$= \frac{\lim_{n \to \infty} \sum_{C(n):(q,v) \in C(n)} P^\infty(C(n))}{\lim_{n \to \infty} \sum_{v'} \sum_{C(n):(q,v') \in C(n)} P^\infty(C(n))} \tag{10.3}$$

is well-defined for all q that have nonzero probability of being in a claims set generated under the claims distribution. We refer to this conditional distribution $\overline{P}(v \mid q)$ as the **answer distribution** of the NDR machine.[4] We will sometimes abuse terminology and use the same expression, "answer distribution", even if we are implicitly

---

[2] This is equivalent to requiring that an NDR machine is a "sequential information source" [8]. In the current context, it imposes restrictions on how likely the NDR machine is to remove claims from the claims tape.

[3] Note that even if a claims set C is small, it might only arise with non-negligible probability in large claims lists, i.e., claims lists produced after many iterations of the NDR machine. For example, this might happen in the NDR machine of the community of mathematicians if the claims in c would not even make sense to mathematicians until the community of mathematicians has been investigating mathematics for a long time.

[4] Note the implicit convention that $\overline{P}(v \mid q)$ concerns the probability of a claims list containing a single claim in which the answer v arises for the precise question q, *not* the probability of a claims

considering $\overline{P}(v \mid q)$ restricted to a proper subset of the questions q that can be produced by the NDR machine. As shorthand we will sometimes write answer distributions as $\mathcal{A}$.

A **mistake-free answer distribution** is one that can be produced by some mistake-free NDR machine. In general, there are an infinite number of NDR machines that all result in the same answer distribution $\mathcal{A}$. However, all NDR machines that result in a mistake-free answer distribution must themselves be mistake-free. For any claims list C and question q such that $\overline{P}(q, C) \neq 0$, we define

$$\overline{P}(v \mid q, C) := \frac{\overline{P}((q, v), C)}{\overline{P}(q, C)} \tag{10.4}$$

$$:= \frac{\lim_{n \to \infty} \sum_{C(n):(q,v) \cup C \in C(n)} P^\infty(C(n))}{\lim_{n \to \infty} \sum_{C(n), v':(q,v') \cup C \in C(n)} P^\infty(C(n))} \tag{10.5}$$

and refer to this as a **generalized** answer distribution. (In the special case that C is empty, the generalized answer distribution reduces to the answer distribution defined in (10.3).)

Claims distributions and (generalized) answer distributions are both defined in terms of the stochastic process that begins with the PTM's question and claims tapes in their initial, blank states. We make analogous definitions conditioned on the PTM having run long enough to have produced a particular claims list C at some iteration. (This will allow us to analyze the far-future distribution of claims of the actual current community of human mathematicians, conditioned on the actual claims list C that that community has currently produced.)

Paralleling the definitions above, choose any pair $n_1, n_2 > n_1$ and any $C_{n_1} \in \mathcal{C}_{n_1}$ such that there is nonzero probability that the NDR machine will produce a sequence of claims lists one of which is C_{n_1}. We add the requirement that the probability distribution of sequences of claims $C^k(n_2) \in \mathcal{C}_{n_2}$ at the beginning of the claims list C that has been produced by the NDR machine at its k'th iteration after starting from its initial state, conditioned on its having had the claims list C_{n_1} on its claims tape at some iteration $< k$, converges in probability in the limit of $k \to \infty$. With abuse of notation, we write that probability distribution as $P^\infty_{C_{n_1}}(C(n_2))$, and require that $P^\infty_{C_{n_1}}(C(n_2))$ is given by marginalizing out the last claim in $P^\infty_{C_{n_1}}(C(n_2 + 1))$. This distribution defines a probability distribution over all (unordered) claims sets $c = \{c_i : i = 1, \ldots, m\}$ containing $m \leq n_2$ claims:

$$P^\infty_{C_{n_1};n_2}(c) := \sum_{C(n_2):\forall i, c_i \in C(n_2)} P^\infty_{C_{n_1}}(C(n_2)) \tag{10.6}$$

We assume that $\lim_{n_2 \to \infty} P^\infty_{C_{n_1};n_2}(c)$ is well-defined for any finite claims set c (for all C_{n-1} that are produced by the NDR machine with nonzero probability). We refer to

---

list that has an answer v in some claim, and that also has the question q in some (perhaps different) claim.

this as a **list-conditioned** claims distribution, for conditioning claims list C_{n_1}, and write it as $\overline{P}_{C_{n_1}}(c)$. It defines an associated **list-conditioned** answer distribution, which we write as $\mathcal{A}_{C_{n_1}}(v \mid q) = \overline{P}_{C_{n_1}}(v \mid q)$. We define the list-conditioned generalized answer distribution analogously. Intuitively, these are simply the distributions over bodies of mathematics that might be produced by the far-future community of mathematicians, conditioned on their having produced the claims list C_{n_1} sometime in their past, while they were still young.

Note that the generalized answer distribution $\overline{P}(v \mid q, c)$ is defined in terms of a claims set c which might have probability zero of being a contiguous sequence of claims, i.e., a claims list. In contrast, $\overline{P}_{C_{n_1}}(v \mid q)$ is defined in terms of a contiguous claims list C_{n_1}. Moreover, the claims in C_{n_1} might have zero probability under the claims distribution, e.g., if the NDR machine removes them from the claims tape during the iterations after it first put them all onto the claims tape. Finally, note that both $\overline{P}(v \mid q, \{c\})$ and $\overline{P}_{C_{n_1}}(v \mid q)$ are limiting distributions of the final conclusions of the far-future community of mathematicians. Both of these differ from the probability that as the NDR machine governing the current community of mathematicians evolves, starting from a current claims list and with a current open question q, it generates the answer v for that question. (That answer might get overturned by the far-future community of mathematicians.)

10.4 Connections to Actual Mathematical Practice

In this section we show how NDR machines can be used to quantify and investigate some of the specific features of the behavior of human mathematicians (see also [21]). Most of the analysis in this section holds even if we restrict attention to NDR machines whose answer distribution $\mathcal{A}$ is a probabilistic mixture of single-valued functions from $q \rightarrow v$. Intuitively, such NDR machines model scenarios where each question $(\mathcal{S}, \varphi)$ is mapped to a unique valence, but we are uncertain what that map from questions to valences is.

10.4.1 Generating New Research Questions

Given our supposition that the community of practicing mathematicians can be modeled as an NDR machine, what is the precise stochastic process that that NDR machine uses in each iteration, in the step where it adds new questions to Q. Phrased differently, what are the goals that guide how the community of mathematicians decides which open questions to investigate at any given moment?

This is obviously an extremely complicated issue, ultimately involving elements of sociology and human psychology. Nonetheless, it is possible to make some high-level comments. First, most obviously, one goal of human mathematicians is that there be high probability that they generate questions whose valence is either t, a or u. Human mathematicians don't want to "waste their time" considering questions (S, φ) where it turns out that φ is not a WFF under S. So we would expect there to be low probability that any such question is added to Q. Another goal is that mathematicians prefer to consider questions whose answer would be a "breakthrough", leading to many fruitful "insights". One way to formalize this second goal is that human mathematicians want to add questions q to Q such that, if they were able to answer q (i.e., if they could determine the valence v of q), then after they did so, and C was augmented with that question-answer pair, the NDR machine would rapidly produce answers to many of the *other* open questions $q \in Q$.

10.4.2 Bayesian Models of Heuristics of Human Mathematicians—General Considerations

Human mathematicians seem to act somewhat like Bayesian learners; as mathematicians learn more by investigating open mathematical questions—as their data set of mathematical conclusions grows larger—they update their probability distributions over those open questions. For example, modern computer scientists assign a greater probability to the claim NP $\neq$ P than did computer scientists of several decades ago. In the remainder of this section we show how to model this behavior in terms of NDR machines, and thereby gain new perspectives on some of the heuristic rules that seem to govern the reasoning of the human mathematical community

First, note that the subjective relative beliefs of the current community of mathematicians do not arise in the analysis up to this point. All probability distributions considered above concern what answers mathematicians are in fact likely to make, as the physical universe containing them evolves, not the answers that mathematicians happen to currently believe. Rather than introduce extra notation to explicitly model the current beliefs of mathematicians, for simplicity we suppose that the subjective relative beliefs of the current community of mathematicians with respect to what the answer is to all questions in the current questions tape, matches the actual answer distribution of the far-future community of mathematicians. As an example, under this supposition, if C is the current claims list of the community of mathematicians and ϕ is the WFF "NP $\neq$ P" phrased in some particular formal system S, then the current relative beliefs of the community of mathematicians concerning whether NP $\neq$ P just equals $\overline{P}(v = t \mid \phi, S)$.[5]

---

[5] In general, even if a mathematician updates their beliefs in a Bayesian manner, the priors and likelihoods they use to do so may be "wrong", in the sense that they differ from the ones used by the far-future community of mathematicians. The use of purely Bayesian reasoning, by itself, provides no advantage over using non-Bayesian reasoning—unless the subjective priors and likelihoods of the current community of mathematicians happen to agree with those of the far-future community of mathematicians. In the rest of this section we assume that there is such agreement. See [5, 23] for how to analyze expected performance of a Bayesian decision-maker once we allow for the

10.4.3 A Bayesian Justification of Abduction in Mathematical Reasoning

Adopting this perspective, it is easy to show that the heuristic technique of "abductive reasoning" commonly used by human mathematicians is Bayes rational. To begin, let $q = (S, \varphi), q' = (S, \varphi')$ be two distinct open questions which share the same formal system S and are both contained in the current set of open questions of the community of mathematicians, Q, and so neither of which are contained in the current claims list of the community of mathematicians, C. Suppose as well that both q and q' occur in $\overline{P}_C$ with probability 1, i.e., the far-future community of mathematicians definitely has answers to both questions. Suppose as well that if q' were a theorem under S, that would make it more likely that q was also a theorem, i.e., suppose that

$$\overline{P}_C\left(v = t \mid q, (q', t)\right) > \overline{P}_C\left(v = t \mid q\right) \tag{10.7}$$

i.e.,

$$\frac{\overline{P}_C\left((q, t), (q', t)\right)}{\overline{P}_C\left(q, (q', t)\right)} > \frac{\overline{P}_C\left((q, t)\right)}{\overline{P}_C(q)} \tag{10.8}$$

and so repeatedly using our assumption that both q and q' occur with probability 1,

$$\frac{\overline{P}_C\left((q, t), (q', t)\right)}{\overline{P}_C\left((q', t)\right)} > \overline{P}_C\left((q, t)\right) \tag{10.9}$$

$$\frac{\overline{P}_C\left((q, t), (q', t)\right)}{\overline{P}_C\left((q, t)\right)} > \overline{P}_C\left((q', t)\right) \tag{10.10}$$

$$\frac{\overline{P}_C\left((q, t), (q', t)\right)}{\overline{P}_C\left(q', (q, t)\right)} > \frac{\overline{P}_C\left((q', t)\right)}{\overline{P}_C(q')} \tag{10.11}$$

i.e.,

$$\overline{P}_C\left(v = t \mid q', (q, t)\right) > \overline{P}_C\left(v = t \mid q'\right) \tag{10.12}$$

So *no matter what the (list-conditioned, generalized) answer distribution of the far-future community of mathematicians $\overline{P}_C$ is*, the probability that q' is true goes up if q is true. Therefore under our supposition that the subjective beliefs of the current community of mathematicians are given by the claims distribution $\overline{P}_C$, not only is it Bayes-rational for them to increase their belief that q' is true if they find that q is—modifying their beliefs this way will also lead them to mathematical truths (if we

---

possibility that the priors they use to make decisions differ from the real-world priors that determine the expected loss of their decision-making.

define "mathematical truths" by the claims distribution of the far-future community of mathematicians).[6]

Stripped down, this inference pattern can be explained in two simple steps. First, suppose that mathematicians believe that some hypothesis H would be more likely to be true if a different hypothesis H' were true. Then if they find out that H actually is true, they must assign higher probability to H' also being true. This general pattern of reasoning, in which we adopt a greater degree of belief in one hypothesis because it would lend credence to some other hypothesis that we already believe to be true, is known as "abduction" [13], and plays a prominent role in actual mathematical practice [21]. As we have just shown, it is exactly the kind of reasoning one would expect mathematicians to use if they were Bayesian reasoners making inferences about their own answer distribution $\mathcal{A}$.

10.4.4   A Bayesian Formulation of the Value of Multiple Proof Paths in Mathematical Reasoning

Real human mathematicians often have higher confidence that some question q is a theorem if many independent paths of reasoning suggest that it is a theorem. To understand why this might be Bayes-rational, as before, let C be the current claims list of the community of mathematicians and let Q be the current list of open questions. Let $\{\{c\}_1, \ldots \{c\}_n\}$ be a set of sets of claims, none of which are in C. By Bayes' theorem,

$$\overline{P}_C\big(v = t \mid q, \{c\}_1, \ldots, \{c\}_n\big) = \frac{\overline{P}_C\big(\{c\}_1, \ldots, \{c\}_n \mid (q, t)\big)\overline{P}_C(v = t \mid q)}{\overline{P}_C\big(\{c\}_1, \ldots, \{c\}_n \mid q\big)} \tag{10.13}$$

Expanding $\overline{P}_C\big(\{c\}_1, \ldots, \{c\}_n \mid q\big)$ in the denominator gives

$$\overline{P}_C(v = t \mid q, \{c\}_1, \ldots, \{c\}_n)$$
$$= \frac{\overline{P}_C(\{c\}_1, \ldots, \{c\}_n \mid (q, v = t))\overline{P}_C(v = t \mid q)}{\overline{P}_C(\{c\}_1, \ldots, \{c\}_n \mid (q, v \neq t))\overline{P}_C(v \neq t \mid q) + \overline{P}_C(\{c\}_1, \ldots, \{c\}_n \mid (q, v = t))\overline{P}_C(v = t \mid q)} \tag{10.14}$$

Next, for all $1 < i \leq n$ define

---

[6] Note that this argument doesn't require the answer distribution of the far-future community of mathematicians to be mistake-free. (The possibility that "correct" mathematics contains inconsistencies with some nonzero probability is discussed below, in Sect. 10.5.) Note also that the simple algebra leading from Eq. (10.7) to Eq. (10.12) would still hold even if q and/or q' were not currently an open question, and in particular even if one or both of them were in the current claims list C. However, in that case, the conclusion of the argument would not concern the process of abduction narrowly construed, since the conclusion would also involve the probability that the far-future community of mathematicians overturns claims that are accepted by the current community of mathematicians.

$$\alpha_i := \frac{\overline{P}_C(\{c\}_1, \ldots, \{c\}_i \mid (q, v = t))}{\overline{P}_C(\{c\}_1, \ldots, \{c\}_{i-1} \mid (q, v = t))} \tag{10.15}$$

$$= \overline{P}_C(\{c\}_i \mid \{c\}_1, \ldots, \{c\}_{i-1}, (q, v = t)) \tag{10.16}$$

$$\beta_i := \frac{\overline{P}_C(\{c\}_1, \ldots, \{c\}_i \mid (q, v \neq t))}{\overline{P}_C(\{c\}_1, \ldots, \{c\}_{i-1} \mid (q, v \neq t))} \tag{10.17}$$

$$= \overline{P}_C(\{c\}_i \mid \{c\}_1, \ldots, \{c\}_{i-1}, (q, v \neq t)) \tag{10.18}$$

Note that due to Eqs. (10.16) and (10.18), we can write

$$\frac{\alpha_i}{\beta_i} = \frac{\overline{P}_C(v = t \mid q, \{c\}_1, \ldots, \{c\}_{i-1})}{\overline{P}_C(v \neq t \mid q, \{c\}_1, \ldots, \{c\}_{i-1})} \tag{10.19}$$

So $\alpha_i \geq \beta_i$ iff $\overline{P}_C(v = t \mid q, \{c\}_1, \ldots, \{c\}_{i-1}) \geq 1/2$. We say that all $\{c\}_i$ in the set $\{\{c\}_i\}$ are **proof paths** if $\alpha_i \geq \beta_i$ for all $1 < i \leq n$.

As an example, suppose that in fact for all $1 < i \leq n$,

$$\overline{P}_C(\{c\}_i \mid \{c\}_1, \ldots, \{c\}_{i-1}, (q, v = t)) = \overline{P}_C(\{c\}_i \mid (q, v = t)) \tag{10.20}$$

$$\overline{P}_C(\{c\}_i \mid \{c\}_1, \ldots, \{c\}_{i-1}, (q, v \neq t)) = \overline{P}_C(\{c\}_i \mid (q, v \neq t)) \tag{10.21}$$

In this case, $\{c\}_i$ is a proof path so long as the probability that the far-future community of mathematicians concludes the claims in $\{c\}_i$ are all true is larger if they also conclude that q is true than it is if they conclude that q is not true. Intuitively, if the claims in $\{c\}_i$ are more likely to lead to the conclusion that q is true (i.e., are more likely to be associated with the claim (q, t)) than to the conclusion that q is false, then $\{c\}_i$ is a proof path.

Plugging Eqs. (10.15) and (10.18) into (10.14) gives

$$\overline{P}_C(v = t \mid q, \{c\}_1, \ldots, \{c\}_n)$$

$$= \frac{\alpha_n}{\beta_n} \frac{\overline{P}_C(\{c\}_1, \ldots, \{c\}_{n-1} \mid (q, v = t))\overline{P}_C(v = t \mid q)}{\overline{P}_C(\{c\}_1, \ldots, \{c\}_{n-1} \mid (q, v \neq t))\overline{P}_C(v \neq t \mid q) + \frac{\alpha_n}{\beta_n}\overline{P}_C(\{c\}_1, \ldots, \{c\}_{n-1} \mid (q, v = t))\overline{P}_C(v = t \mid q)} \tag{10.22}$$

If we evaluate Eq. (10.14) for $n - 1$ rather than n and then rearrange it to evaluate the numerator in Eq. (10.22), we get

$$\frac{\overline{P}_C(v = t \mid q, \{c\}_1, \ldots, \{c\}_n)}{\overline{P}_C(v = t \mid q, \{c\}_1, \ldots, \{c\}_{n-1})}$$

$$= \frac{\alpha_n}{\beta_n} \frac{\overline{P}_C(\{c\}_1, \ldots, \{c\}_{n-1} \mid (q, v \neq t))\overline{P}_C(v \neq t \mid q) + \overline{P}_C(\{c\}_1, \ldots, \{c\}_{n-1} \mid (q, v = t))\overline{P}_C(v = t \mid q)}{\overline{P}_C(\{c\}_1, \ldots, \{c\}_{n-1} \mid (q, v \neq t))\overline{P}_C(v \neq t \mid q) + \frac{\alpha_n}{\beta_n}\overline{P}_C(\{c\}_1, \ldots, \{c\}_{n-1} \mid (q, v = t))\overline{P}_C(v = t \mid q)} \tag{10.23}$$

$$:= \epsilon_n \tag{10.24}$$

Iterating gives

$$\overline{P}_C\big(v = t \mid q, \{c\}_1, \ldots, \{c\}_n\big) = \overline{P}_C\big(v = t \mid q, \{c\}_1\big) \prod_{i=2}^{n} \epsilon_i \qquad (10.25)$$

Next, note that $\alpha_i \geq \beta_i$ implies that $\epsilon_i \geq 1$. So Eq. (10.25) tells us that if each $\{c\}_i$ is a proof path, i.e., $\epsilon_i > 1$ for all $i > 1$, then the posterior probability of q being true keeps growing as more of the n proof paths are added to the set of claims accepted by the far-future community of mathematicians.

This formally establishes the claim in the introduction, that the NDR machine model of human mathematicians lends formal justification to the idea that, everything else being equal, a mathematical claim should be believed more if there are multiple distinct lines of reasoning supporting that claim.

10.5 Measures over Multiverses

The mathematical universe hypothesis (MUH) argues that our physical universe is just one particular formal system, namely, the one that expresses the laws of physics of our universe [11, 15, 17–20]. Similar ideas are advocated by Barrow [3, 4], who uses the phrase "pi in the sky" to describe this view. Somewhat more precisely, the MUH is the hypothesis that our physical world is isomorphic to a formal system. A key advantage of the MUH is that it allows for a straightforward explanation of why it is the case that, to use Wigner's [22] phrase, mathematics is "unreasonably effective" in describing the natural world. If the natural world is, by definition, isomorphic to mathematical structures, then the isometry between nature and mathematics is no mystery; rather, it is a tautology. While the MUH is accepted (implicitly or otherwise) by many theoretical physicists working in cosmology, some disagree with various aspects of it; for an overview of the controversy, see Hut et al. [11].

Here, we adapt the MUH into the framework of NDR machines. Suppose we have a claims distribution that is a delta function about some formal system S, in the sense that the probability of any claim whose question does not specify the formal system S is zero under that distribution. Similarly, suppose that any string φ which is a WFF under S has nonzero probability under the claims distribution. (The reason for this second condition is to ensure that the answer distribution, $\mathcal{A}(v \mid (S, \varphi)) = \overline{P}(v \mid (S, \varphi))$, is well-defined for any φ which is a WFF under S.) We refer to the associated pair $(S, \mathcal{A})$ of any such claims distribution as an **NDR world**. Similarly, we define an **NDR world instance** of an NDR machine as any associated pair (S, c) where c is a maximal claims set of that NDR machine.

Intuitively, an NDR world is the combination of a formal system and the set of answers that some NDR machine would provide to questions formulated in terms of that formal system, without specifying a distribution over such questions. An NDR world instance is a sample of that NDR machine. (It is not clear what a distribution over questions would amount to in a physical universe, which is why we exclude such distributions from both definitions.) A mistake-free NDR world is any NDR world with a mistake-free answer distribution, and similarly for an NDR world instance. Note that while a mistake-free NDR world can only produce an NDR world instance that is mistake-free, mistake-free NDR world instances can be produced by NDR worlds that are not mistake free.

Rephrased in terms of our framework, previous versions of the MUH hold that our physical universe is a mistake-free NDR world. That is, the physical universe is isomorphic to a particular formal system S which in turn assigns, with certainty, a specific syntactic valence to each possible string in the alphabet of S. Our approach can be used to generalize this in two ways. First, it allows for the possibility that the physical world is isomorphic to an NDR world that is not mistake-free. Second, it allows for the possibility that the physical world is isomorphic to an NDR world *instance* that is not mistake-free. In such a world, some strings would have their syntactic valence not because of the perfect application of the rules of some formal system, but rather because of the stochastic application of those very rules.

Thus, our augmented version of the MUH allows for the possibility that *mathematical* reality is fundamentally stochastic. So in particular, the mathematical reality governing our physical universe may be stochastic. This is similar to the fact that *physical* reality is fundamentally stochastic (or at least can be interpreted that way, under some interpretations of quantum mechanics).

An idea closely related to the MUH as just defined is the mathematical multiverse hypothesis (MMH). The MMH says that some non-singleton subset of formal systems is such that there is a physical universe that is isomorphic to each element of that subset. Each of these possible physical universes is taken to be perfectly *real*, in the sense that the formal system to which that universe is isomorphic is not just the fictitious invention of a mathematician, but rather a description of a physical universe. In this view, the world that we happen to live in is unique not because it is uniquely real, but because it is our *actual* world. Following Lewis [12], defenders of the MMH understand claims about 'the actual universe' as *indexical* expressions, i.e. expressions whose meaning can shift depending on contingent properties of their speaker (pp. 85–86).

A central concern of people working on the MMH (e.g. Schmidhuber [15] and Tegmark [20]) is how to specify a probability measure over the set of all universes, which we will refer to as an **MMH measure**. Implicitly, the concern is not merely to specify the subjective degree of belief of us humans about what the laws of physics are in our particular universe. (After all, the MMH measures considered in the literature assign nonzero prior probability to formal systems that are radically different from the laws of our universe, supposing such formal systems are just as "real" as the one that governs our universe.) Rather the MMH measure is typically treated as more akin to the objective probability probability distributions that arise in quantum mechanics, as quantifying something about reality, not just about human ignorance.

In existing approaches to MMH measures, it is assumed that any physical reality is completely described by a set of recursive rules that assign, with certainty, a particular syntactic valence to any string. As mentioned above, this amounts to the assumption that all physical universes are mistake-free NDR worlds. So the conventional conception of an MMH measure is a distribution over mistake-free NDR worlds, i.e., over NDR world instances that are mistake-free. A natural extension, of course, is to have the MMH measure be a distribution over *all* NDR world instances, not just those that are mistake-free. A variant would be to have the MMH measure be a distribution over all NDR worlds, not just those that are mistake-free. Another possibility, in some ways more elegant than these two, would be to use a single NDR machine to define a measure over NDR world instances, and identify that measure as the MMH measure.

10.6 Future Research Directions

There are many possible directions for future research. For example, in general, for any q produced with nonzero probability, the PTM underlying the NDR machine of the community of mathematicians will cause the answer distribution $\overline{P}(v \mid q)$ not to be a delta function about one particular valence v. This is also true for distributions concerning the current community of mathematicians: letting $((q, v), C)$ be the current sequence of claims actually accepted by that community, and supposing it was produced by k iterations of the underlying PTM, $P^k(v \mid q, C)$ need not equal 1. In other words, if we were to re-sample the stochastic process that resulted in the current claims list of the community of mathematicians, then even conditioning on the question q being on that claims list, and even conditioning on the *other*, earlier claims in that list, C, there may be nonzero probability of producing a different answer to q from the one actually accepted by the current community of mathematicians.

This raises the obvious question of how $P^k(v \mid q, C)$ would change if we modified the update distribution of the PTM underlying that NDR machine. In particular, there are many famous results in the foundations of mathematics that caused dismay when they were discovered, starting with the problems that were found in naive set theory, through Godel's incompleteness theorems on to the proof that both the continuum hypothesis and its negation are consistent with the axioms of modern set theory. A common feature of these mathematical results is that they restrict mathematics itself, in some sense, and so have implications for the answers to many questions. Note though that all of those results were derived using deductive reasoning, expressible in terms of a formal system. So they can be formulated as claims by an NDR machine. This raises the question of how robust those results are with respect to the noise level in that NDR machine. More precisely, if those results are formulated as claims of the NDR machine, and some extremely small extra stochasticity is introduced into the PTM underlying the NDR machine, do the probabilities of those results— the probability distribution over the valences associated with the questions—change radically? Can we show that the far-ranging results in mathematics that restrict its own capabilities are fragile with respect to errors in mathematical reasoning? Or conversely, can we show that they are unusually robust with respect to such errors?

As another example of possible future research, the field of epistemic logic is concerned with how to formally model what it means to "know" that a proposition is true. Most epistemic logic models require that knowledge be *transitive*, meaning that if one knows some proposition A, and knows that $A \Rightarrow B$, then one knows B [1, 6]. Such models are subject to an infamous problem known as **logical omniscience**: supposing only that one knows the axioms of standard number theory and Boolean algebra, by recursively applying transitivity it follows that one "knows" *all* theorems in number theory—which is clearly preposterous.

Note though that any such combination of standard number theory and Boolean algebra is a formal system. This suggests that we replace conventional epistemic logic with an NDR machine version of epistemic logic, where the laws of Boolean algebra are only stochastic rather than iron-clad. In particular, by doing that, the problem of logical omniscience may be resolved: it may be that for any non-zero level of noise in the NDR machine, and any $0 < \epsilon < 1$, there is some associated finite integer n such that one knows no more than n theorems of number theory with probability greater than ϵ.

As another example of possible future work, the models of practicing mathematicians as NDR machines introduced in Sect. 10.4 are very similar to the kind of models that arise in active learning [16], a subfield of machine learning. Both kinds of model are concerned with an iterated process in which one takes a current data set C, consisting of pairs of inputs (resp., questions) and associated outputs (resp., valences); uses C to suggest new inputs (resp., questions); evaluates the output (resp., valence) for that new input (resp., question); and adds the resulting pair to the data set C. This formal correspondence suggests that it may be fruitful to compare how modern mathematical research is conducted with the machine learning techniques that have been applied to active learning, etc.

In this regard, recall that the *no free lunch theorems* are a set of formal bounds on how well any machine learning algorithm or search algorithm can be guaranteed to perform if one does not make assumptions for the prior probability distribution of the underlying stochastic process [23, 25]. Similar bounds should apply to active learning. Given the formal correspondence between the model of mathematicians as NDR machines and active learning algorithms, this suggests that some version of the NFL theorems should be applicable to the entire enterprise of mathematics research. Such bounds would limit how strong any performance guarantees for modern mathematical research practices can be without making assumptions for the prior distribution over the possible answer distributions of the infinite-future community of mathematicians, $\overline{P}(\mathcal{A})$.

10.7 Conclusion

Starting from the discovery of non-Euclidean geometry, mathematics has been greatly enriched whenever it has weakened its assumptions and expanded the range of formal possibilities that it considers. Following in that spirit of weakening assumptions,

we introduced a way to formalize mathematics in a stochastic fashion, without the
the assumption that mathematics itself is fully deterministic. We showed that this
formalism justifies some common heuristics of actual mathematical practice. We also
showed how it extends and clarifies some aspects of the multi-universe hypothesis.

A Probabilistic Turing Machines

Perhaps the most famous class of computational machines are Turing machines. One
reason for their fame is that it seems one can model any computational machine that
is constructable by humans as a Turing machine. A bit more formally, the **Church-
Turing thesis** states that "a function on the natural numbers is computable by a
human being following an algorithm, ignoring resource limitations, if and only if it
is computable by a Turing machine."

There are many different definitions of Turing machines (TMs) that are "computationally equivalent" to one another. For us, it will suffice to define a TM as a
7-tuple $(R, \Lambda, b, v, r^{\varnothing}, r^A, \rho)$ where:

1. R is a finite set of **computational states**;
2. Λ is a finite **alphabet** containing at least three symbols;
3. $b \in \Lambda$ is a special **blank** symbol;
4. $v \in \mathbb{Z}$ is a **pointer**;
5. $r^{\varnothing} \in R$ is the **start state**;
6. $r^A \in R$ is the **halt state**; and
7. $\rho : R \times \mathbb{Z} \times \Lambda^{\infty} \to R \times \mathbb{Z} \times \Lambda^{\infty}$ is the **update function**. It is required that for
   all triples (r, v, T), that if we write $(r', v', T') = \rho(r, v, T)$, then v' does not
   differ by more than 1 from v, and the vector T' is identical to the vectors T for
   all components with the possible exception of the component with index v^7;

We sometimes refer to R as the states of the "head" of the TM, and refer to the third
argument of ρ as a **tape**, writing a value of the tape (i.e., of the semi-infinite string
of elements of the alphabet) as T.

Any TM $(R, \Sigma, b, v, r^{\varnothing}, r^A, \rho)$ starts with $r = r^{\varnothing}$, the counter set to a specific
initial value (e.g, 0), and with T consisting of a finite contiguous set of non-blank
symbols, with all other symbols equal to b. The TM operates by iteratively applying
ρ, until the computational state falls in r^A, at which time it stops, i.e., any ID with
the head in the halt state is a fixed point of ρ.

If running a TM on a given initial state of the tape results in the TM eventually
halting, the largest blank-delimited string that contains the position of the pointer
when the TM halts is called the TM's **output**. The initial state of T (excluding the
blanks) is sometimes called the associated **input**, or **program**. (However, the reader

---

[7] Technically the update function only needs to be defined on the "finitary" subset of $\mathbb{R} \times \mathbb{Z} \times \Lambda^{\infty}$,
namely, those elements of $\mathbb{R} \times \mathbb{Z} \times \Lambda^{\infty}$ for which the tape contents has a non-blank value in only
finitely many positions.

should be warned that the term "program" has been used by some physicists to mean specifically the shortest input to a TM that results in it computing a given output.) We also say that the TM **computes** an output from an input. In general, there will be inputs for which the TM never halts. The set of all those inputs to a TM that cause it to eventually halt is called its **halting set**.

The set of triples that are possible arguments to the update function of a given TM are sometimes called the set of **instantaneous descriptions** (IDs) of the TM. Note that as an alternative to the definition in (7) above, we could define the update function of any TM as a map over an associated space of IDs.

In one particularly popular variant of this definition of TMs the single tape is replaced by multiple tapes. Typically one of those tapes contains the input, one contains the TM's output (if and) when the TM halts, and there are one or more intermediate "work tapes" that are in essence used as scratch pads. The advantage of using this more complicated variant of TMs is that it is often easier to prove theorems for such machines than for single-tape TMs. However, there is no difference in their computational power. More precisely, one can transform any single-tape TM into an equivalent multi-tape TM (i.e., one that computes the same partial function), as shown by Arora and Barak [2].

A **universal Turing machine** (UTM), M, is one that can be used to emulate any other TM. More precisely, in terms of the single-tape variant of TMs, a UTM M has the property that for any other TM M', there is an invertible map f from the set of possible states of the tape of M' into the set of possible states of the tape of M, such that if we:

1. apply f to an input string σ' of M' to fix an input string σ of M;
2. run M on σ until it halts;
3. apply f^{-1} to the resultant output of M;

then we get exactly the output computed by M' if it is run directly on σ'.

An important theorem of computer science is that there exist universal TMs (UTMs). Intuitively, this just means that there exists programming languages which are "universal", in that we can use them to implement any desired program in any other language, after appropriate translation of that program from that other language. The physical CT thesis considers UTMs, and we implicitly restrict attention to them as well.

Suppose we have two strings s^1 and s^2 where s^1 is a proper prefix of s^2. If we run the TM on s^1, it can detect when it gets to the end of its input, by noting that the following symbol on the tape is a blank. Therefore, it can behave differently after having reached the end of s^1 from how it behaves when it reaches the end of the first $\ell(s^1)$ bits in s^2. As a result, it may be that both of those input strings are in its halting set, but result in different outputs. A **prefix (free) TM** is one in which this can never happen: there is no string in its halting set that is a proper prefix of another string in its halting set. For technical reasons, it is conventional in the physics literature to focus on prefix TMs, and we do so here.

The **coin-flipping distribution** of a prefix TM M is the probability distribution over the strings in M's halting set generated by IID "tossing a coin" to generate those strings, in a Bernoulli process, and then normalizing. So any string σ in the halting set has probability $2^{-|\sigma|}/\Omega$ under the coin-flipping prior, where Ω is the normalization constant for the TM in question.

Finally, for our purposes, a **Probabilistic Turing Machine** (PTM) is a conventional TM as defined by conditions (1)–(7), except that the update function ρ is generalized to be a conditional distribution. The conditional distribution is not arbitrary however. In particular, we typically require that there is zero probability that applying such an update conditional distribution violates condition (7). Depending on how we use a PTM to model NDR machines, we may introduce other requirements as well.

References

1. S. Aaronson, Why philosophers should care about computational complexity, in *Computability: Turing, Gödel, Church, and Beyond*, pp. 261–327 (MIT Press, 2013)
2. S. Arora, B. Barak, *Computational Complexity: A Modern Approach* (Cambridge University Press, 2009)
3. J.D. Barrow, *Theories of Everything: The Quest for Ultimate Explanation* (Clarendon Press, Oxford, 1991)
4. J.D. Barrow, Godel and physics. *Kurt Gödel and the Foundations of Mathematics: Horizons of Truth*, p. 255 (2011)
5. J.L. Carroll, A Bayesian decision theoretical approach to supervised learning, selective sampling, and empirical function optimization (2010)
6. R. Fagin, Y. Moses, J.Y. Halpern, M.Y. Vardi, *Reasoning About Knowledge* (MIT Press, 2003)
7. K. Gödel, *On Undecidable Propositions of Formal Mathematics Systems* (Institute for Advanced Study, 1934)
8. P. Grunwald, P. Vitányi, Shannon information and Kolmogorov complexity. arXiv preprint arXiv:cs/0410002 (2004)
9. D. Hilbert, Die grundlagen der mathematik, in *Die Grundlagen der Mathematik*, pp. 1–21 (Springer, 1928)
10. D. Hume, *A Treatise of Human Nature* (Courier Corporation, 2012). Book 1, Part 4, Section 1
11. P. Hut, M. Alford, M. Tegmark, On math, matter and mind. Found. Phys. **36**(6), 765–794 (2006)
12. D. Lewis, *Counterfactuals* (Basil Blackwell, Oxford, 1973)
13. C.S. Peirce, *Collected Papers of Charles Sanders Peirce*, vol. 2 (Harvard University Press, 1960)
14. H. Poincaré, Mathematical creation. The Monist 321–335 (1910)
15. J. Schmidhuber, A computer scientist's view of life, the universe, and everything, in *Foundations of Computer Science*, pp. 201–208 (Springer, 1997)
16. B. Settles, Active learning literature survey. Technical report, University of Wisconsin-Madison Department of Computer Sciences (2009)
17. M. Tegmark, Is "the theory of everything" merely the ultimate ensemble theory? Ann. Phys. **270**(1), 1–51 (1998)
18. M. Tegmark, The mathematical universe. Found. Phys. **38**(2), 101–150 (2008)
19. M. Tegmark, The multiverse hierarchy. arXiv preprint arXiv:0905.1283 (2009)
20. M. Tegmark, *Our Mathematical Universe: My Quest for the Ultimate Nature of Reality* (Vintage, 2014)
21. S. Viteri, S. DeDeo, Explosive proofs of mathematical truths. arXiv preprint arXiv:2004.00055 (2020)

22. E.P. Wigner, *The Unreasonable Effectiveness of Mathematics in the Natural Sciences*, vol. 13, pp. 1–14 (1960)
23. D.H. Wolpert, The lack of a priori distinctions between learning algorithms. Neural Comput. **8**(7), 1341–1390 (1996)
24. D.H. Wolpert, The stochastic thermodynamics of computation. J. Phys. A Math. Theor. **52**(19), 193001 (2019)
25. D.H. Wolpert, W.G. Macready, No free lunch theorems for optimization. IEEE Trans. Evol. Comput. **1**(1), 67–82 (1997)

Chapter 11
Computational Complexity as Anthropic Principle: A Fable

Rick Searle

> Give me a place to stand and with a lever I will move the whole world.
> Archimedes.
> *I am a strange loop.*
> Douglas Hofstadter.

In the early 19th century the polymath and physicist Pierre Simon de Laplace imagined a demon. Laplace posited that a super-intellect in possession of all the data of nature, with the application of a single equation, would possess complete knowledge of not only the past, but the entirety of the future as well. What we have learned in the two centuries since he conceived of his demon is that its existence is almost certainly impossible. The laws of physics themselves prohibit the existence of any such super-intellect. And while that fact, in and of itself, may seem like a trivial observation there may be a way in which it is telling us something deep about the universe in which we live. Rather than merely serve as an ultimate limit to the human pursuit of knowledge, we might instead use the impossibility of Laplacian demons as a tool to infer aspects of nature beyond what we currently know.

However, I will first need to demonstrate how we learned that Laplacian demons are impossible in the first place. Permit me to do so in the most engaging way I know, namely, in the form of a fantastical story.

11.1 Laplace Builds a Demon

For now, he was safe in the tranquility of the countryside. The very moment the rumors of yet another revolution had begun circulating in Paris, Laplace had packed up his family and fled. Who knew if another Terror would unleash its angry mobs upon the citizens of France? Who knew if the reactionaries gathering under the banner of Bonaparte, fresh from his plundering of Egypt, might themselves stage a coup and hurl the country towards an even darker fate, the crowd brought to the point of frenzy

R. Searle (✉)
Institute for Ethics and Emerging Technology (IEET), Boston, MA, USA

by a demagogue and enthroning him as dictator—like they did long ago with Caesar in Rome?

Years earlier he had seen the lives of his friends Condorcet and Lavoisier lost to witch—hunts against imaginary enemies. In light of this he had done everything possible to avoid the debates of *politics* and focused himself exclusively on problems he believed to be free from the foibles of human emotion and judgement—the questions of science. There was no knowing if this indifference might itself eventually lead to him being labeled an enemy by either the revolutionaries or the reactionaries. Might the choice to do nothing itself be deemed a choice? The revolution of the heavenly bodies was deterministic, predictable, clocklike. The revolutions of human beings, however, were stochastic, terrifying, shrouded in perpetual uncertainty. If only he could pierce even part of the veil.

It was during the height of his anxiety that Laplace was given the scroll. Into the hands of a scholar at the Académie des Sciences had come an ancient document from the Ottoman land of Crete that had proven indecipherable to all who had tried to crack its code. It appeared to be a sort of blueprint, but the exact nature or purpose of the machine it described was utterly mysterious. It certainly was written using some form of sophisticated mathematics, though it utilized the geometric reasoning of the Greeks rather than the modern algebra, and for this reason the Académie had turned to Laplace.

At first a mere useless puzzle to secure needed distraction, it was fear that brought forth his epiphany. Startled awake one evening by a loud crack of thunder, Laplace, for once, could remember his dream. The storm must have raged for some time before it awakened him. Asleep, the lightning had echoed in his unconscious mind with visions of entrails, murmurations, fractal scars on oracle bones. Awake, he now realized that the device the scroll described was a sort of predictor—a means of calculating the future based upon a kind of geometric calculus built of branching patterns.

Was this not a matter of great resonance? The ancient Greeks like the moderns had replaced magic with science but found the prospect of divining the future irresistible. Here was their attempt to transform the superstition of the Oracle at Delphi into something real and based on the true laws of nature. Was it not at least a possibility that the blueprint he had before him had emerged from the mind of the great Archimedes himself, that what it described was a kind of lens for peering through the opacity of the natural world and seeing through to the logical structure that lie beneath?

At dawn he set about building the instrument. He sent orders for a multitude of bizarre gears and levers to the finest machinists in Paris. It took him the better part of a year to complete it—punctuated by what proved only temporary frustrations and failures.

When completed the mechanism filled almost a large room, the greatest extent of which was taken up by a large sort of pendulum, in its center a thin needle which gently glanced a circular layer of fine sand. Questions to the oracle were posed by manipulating a series of nobs upon which were inscribed the script of the ancient Greek alphabet with letters used for numbers in the Milesian style. Pulling a lever returned the circle of sand to its prior, pristine state—smooth, symmetrical, featureless—and from the viewpoint of any particular grain of sand—perfectly random.

What perplexed and frustrated him was that the answers to his queries came back in the form of inscrutable scribbles drawn upon the sand by the pendulum. Laplace cursed himself for having done something wrong, but then he noticed the regularity of the drawings.

He had an idea. He took a pair of dice and asked through the dials for a prediction of their outcome before he threw them. The instrument then drew lines in the style of branched tallying. Its predictions always perfectly matching the number of the dice that Laplace would throw. If it drew a branch with three lines followed by a branch with two, his roll would inevitably be a three and a two.

Other experiments came to him. In the habit of skipping morning and afternoon meals as he became lost in his work, his wife would often ask one of the house servants to bring him something to eat during the late afternoon. What would be brought was always a surprise to Laplace, his sustenance depending on what was available in the kitchen, a question which he found to be of no interest. He himself couldn't predict what might be sent to him but could the machine?

"What will Marie Charlotte send me today?" he asked by pulling at the nobs and levers. Upon entering his question, the device drew a horizontal line that branched in two and then moved outward then inward until the lines had formed what were three distinct groups of shapes only to end by joining together and stopping. When the maid appeared sometime later with a plate of bread and cheese along with a small bottle of wine, it stuck Laplace that the shapes in the sand could very well represent the meal he had before him.

Laplace took a slug from the bottle and sat down dizzy with confusion. "Could it be a form of pictograph?" he asked aloud with the maid having left and only the mute contraption in front of him to hear. At once he bolted from his lab and in the direction of his study leaving his food for the rats.

With his copy of the Encyclopédie Méthodique before him Laplace began searching through all the known versions of the world's pictographic scripts. None of the shapes he had seen fit, neither the rectangular contortions of the ancient Chinese script or its many derivatives, nor did any of the other notable variants of picture language such as Cuneiform or Mesoamerican. It was thus in a sense of near hopelessness that he looked at the most infamous, and until very recently, least understood pictographic language of them all—Egyptian hieroglyphs.

To his amazement the drawings did indeed appear to conform to the shapes of well-known hieroglyphs. The ancient Egyptian word for bread composed of a string of pictures—foot, oval, bird, hand—clearly matched the contours of the lines in the sand. The device was writing in the language of the pharaohs.

This was surely the hand of fate for it was at about this time that rumor had it that amidst the horde of treasures the stout Corsican had looted from Egypt was a Ptolemaic stele upon which was inscribed a code for translating hieroglyphics into ancient Greek. Few scholars had had the opportunity to study the object, but a man who had was the Egyptologist Jean-François Champollion, who (again), fortune would have it, had sought refuge from the political turmoil of the capital in the exact same village as Laplace himself.

Laplace commandeered one of his servants to call upon Champollion and bring him to the estate with utmost haste. Within only a few hours the Egyptologist had arrived, his look laying bare his anxiety, for surely he had thought such a call for aid was a harbinger of ill fortune for refugees from political dangers such as himself.

Champollion was quite comforted (perhaps even annoyed) when Laplace explained the reason he had called upon him in such an abrupt manner. That it had nothing to do with Napoleon's coup d'état or the Parisian riots, but with the fabulous device he had built.

Demonstration of the instrument soon made Champollion doubt that what he was seeing was a mere magic trick, though the air of the occult that surrounded its design and powers led him to nickname the machine *the demon*—a term he used alternatively with a tone of derision or fear. A firm believer in the *sapientia antique*, he agreed to serve as a kind of medium for the prognostications of the machine and sent for his notebook on the Egyptian stele so that he could serve as a translator for Laplace.

As the real investigation was about to begin Champollion pleaded with Laplace to ask the demon questions on the future of their beloved France. What would come of the revolution and the power mad reaction that had arisen in its wake? What fate, good or ill, lie before them? All of these entreaties Laplace brushed aside responding almost in the style of a soliloquy on the unpredictability of human affairs as if it were a speech written in advance.

The intractability of human behavior, Laplace lectured, was not a result of individuals being free, for they were no freer than the gears of a clockwork. The problem was that the gears themselves were broken and flawed in an uncountable number of idiosyncratic ways. A working clock is by its very definition predicable. What is man but a broken clock?

If the demon was deemed incapable of predicting the fate of the nation might it not still be capable of discerning his own fate wondered Champollion, and begged Laplace to inquire of the demon whether or not they should flee France for the safety of England.

Something about the tone of Jean-François, or the look of desperation on his face, weakened Laplace's commitment to ask the device scientific questions alone. He had been reminded of his own fear, and tempted by the possibility of assuaging his anxiety by knowing his fate. Later Laplace would recall that the course of what transpired was utterly contingent, one of an uncountable multitude of possible paths into the future that might have been followed instead, but like a play with alternative endings, only one could be chosen.

"Should they flee to England?" Laplace asked by turning the knobs of Greek letters and waited as the device snapped to life and began sketching out its answer in the sand as the Egyptologist looked on in astonishment. Champollion paged frantically through his note book desperate to match the branching lines emerging before him with his drawings of the glyphs from the stele.

"What is its answer?" Laplace eventually barked in impatience.

"I believe it says that it depends on the weather," the Egyptologist responded quizzically.

"If we should fear storms in the channel, then, I'll just ask it to predict the weather." Laplace replied, pulling the lever to clear the sand, and spelling out his question with the nobs. "And what weather does it predict?"

The device slowly drew out another series of branching shapes which Champollion translated using his notes as "incalculable". This shape was followed by what he at first thought was an ankh symbol, but then realized it wasn't a hieroglyphic at all, but a picture of a butterfly.

"Unknowable because of butterflies? I do not know what it means," Champollion blurted out.

"The problem of three bodies!" Laplace exclaimed.

"What?"

"The position of three orbiting bodies cannot be calculated exactly whenever infinitesimal differences in measurement need to be taken into account. The fluttering of a butterfly's wings can give birth to storms."

"There are limits to reason?" Champollion wondered aloud. "Might it at least calculate the best way for us to make our escape, given that the location of the partisans is unknown and always changing?" he asked. At which Laplace again began pulling at the lever and nobs. Once again the oracle responded by drawing out symbols that meant unknowable.

"I have lost myself!", Laplace bellowed. "The answer to our questions depends upon knowledge shrouded in even murkier regions than the invisible molecules determining the weather - the darkness of the human heart. Much better to ask of the instrument questions susceptible to reason alone."

At that moment Champollion thought he could hear the sound of cannons far off in the distance. He said nothing as Laplace appeared to pay no attention. Perhaps he was hearing things. The clanking and whirling of the enormous machine filled him with a sense of *déjà* vu. He remembered a visit to his friend—the great machinists John Joseph Merlin—in the golden days before the time of revolution and war. Merlin displayed for him then his new automata—a gigantic clock, which he claimed ran, and would always run, from the power of the air alone. At the time had not grasped the implications of such an invention thinking it a mere plaything, but now its true import came to him and he said to Laplace.

"I have often wondered how we might have avoided a revolution such as this. What conditions might have prevented the whole of society from crumbling like a castle made of sand? As Rousseau has taught us, it was inequality that caused it. And what causes inequality but the want of surplus in society, a bare minimum left over from basic needs seized upon by those with power. Only a vast increase in the surplus would free us from this state, which might be fulfilled by machines capable of perpetual motion. Ask your demon whether such machines are possible."

At this Laplace entered the question, but instead of beginning its drawing the machine merely circled its pendulum above the sand as if it were unable to make up its mind on where to begin.

Champollion remembered how shocked he was by the scale of the automaton built by Merlin, and how the machinist had claimed that the devise needed to be of such scope were it to run perpetually, and he said to Laplace, "Perhaps it's a matter of size, if the machine were bigger it might answer us quickly or at all?" At which Laplace immediately began to input the question.

"How much bigger would you need to be to answer my question within the span of a day?" The machine drew out its answer.

Champollion again scrambled through his notes, finding the sketch before him nonsensical.

"It's saying its circle would need to have a diameter of over three million river units."

"What is that in kilometers?"

"Ten and one half."

"Damn it, that's over 30,000,000 km!" Laplace, said having quickly ran through the calculation in his head, and then he remembered that he had recently stumbled across a similar number while researching his *Exposition du système du monde*. This figure was 250 times the diameter of the sun, a mass which he had surmised was so great that even light could not escape the pull of its gravity.

A grimace appeared across Laplace's face, and he felt himself unmoored. His mind raced with the perverse, unanticipated realization that the world's form might be the inverse of his prior assumptions. Perhaps it was not just humanity that was stochastic and unpredictable, but, in some deep sense, nature itself. That the child, rather than being flawed, was instead a perfect reflection of its parent—condemned to a terrifying uncertainty.

Laplace felt a wave of anxiety pass over him as he entered his last question. "Even should we never know the future course of events it is comfort enough to know at bottom that they have been determined in advance - that the world is structured according to the laws of some universal intellect far beyond human comprehension. Tell me, oh ancient oracle- is the world at bottom determined or random?" And the machine answered:

"Random."

Laplace felt himself destroyed, as if the entire world he had constructed for himself had proven a lie and was disappearing through his fingers like sand. And then he remembered the paradox of Epimenides he had learned in his youth. "How can this cursed machine know that for all time what we ask of it is unknowable? It is true what they say - all Cretans are liars!"

11.2   The Science that Destroys Demons

In my imagined tale, Laplace's attempt to build a super-intellect is undone by five of the deepest discoveries of science over the last two centuries: deterministic chaos, computational complexity, the Black hole information paradox, and the paradox of self-reference at the heart of both Gödel's Incompleteness Theorem and Turning's notion of uncomputability. I will discuss each in turn.

In the 20th century Edward Lorenz formalized the notion of deterministic chaos, which had been circulating at least since the 19th century genius Henri Poincaré had shown the three body problem could not be solved exactly. "Shadows" of possible trajectories would have to suffice [1]. At bottom the world might be deterministic, but we are unable to predict its exact course for any but the simplest systems. For dynamic and complex phenomenon prediction over long time scales requires infinite precision in terms of data. Seemingly inconsequential differences in measurement have a way of rippling through systems giving rise to radically different outcomes as Lorenz discovered in his investigations of the weather [2].

Yet the limitations on a super-intellect imposed by the need to gather ever more precise data pale in comparison to the question of whether that data can even be processed in the first place. It's here where we encounter the discoveries of computer science, especially since the 1970s [3]. Computational complexity is formally a branch of mathematics, but is perhaps unique among the branches in that it takes the reality of the constraints of time and space seriously.

The problem of finding the best escape route that my imagined Laplace poses to the demon is a variant of what's called "The Canadian Traveler's Problem" (itself a harder version of the infamous "Traveling Salesman Problem.") [4]. It is not that these and other examples of NP complete problems are unsolvable, it is that the time needed to solve them grows exponentially with the size of the problem itself.

Not even quantum computers are expected to be able to solve NP complete problems in polynomial time, let alone problems above that complexity class. The computer scientist Scott Aaronson has gone so far as to propose what he calls 'The NP Hardness Conjecture', that states "There is no physical process to solve NP complete problems in polynomial time" [5].

The implications of computational complexity become extremely important for prediction (and therefore for physics) in the case of black holes and thermodynamics. As Aaronson points out, it's tempting to think we can solve NP complete problems in polynomial time by just throwing more resources at them. We can just use a bigger or faster computer. Yet both of these solutions merely swap one exponential for another. Exponential time for exponential energy. At some point the energy needs get so great that the concentrated mass of the computer exceeds the Schwarzschild Radius and collapses to form a black hole [6].

If the physics of black holes is a unique environment where some combined version of quantum mechanics and general relativity becomes necessary, then for over a generation, physicists have been uncovering clues that such a unification might come in part from applying lessons learned in the field of computational complexity.

Since the work of Jacob Bekenstein we've known that black holes contain the maximum information density. The Bekenstein Bound means that no machine we build will ever have more than a finite amount of memory [7]. As much later pointed out by Patrick Hayden and John Preskill, limits on computation may be a way to solve the famous Black Hole Information Paradox. The information content of a black hole is equally inaccessible both inside and outside the horizon. The inside because of gravity and the outside because the information has become so scrambled [8] it would be impossible within the lifetime of the black hole to decode it [9]. Perhaps not truly random, but certainly pseudo-random [10].

Still we don't need to go to the level of black holes to confront instances where computational complexity plays a deep role in physics. The other famous demon in physics was that of James Clerk Maxwell. Were a Maxwellian demon that could effortlessly separate hot and cold particles in a box possible, we would have created a version of the perpetual motion machine dreamed of since antiquity.

The reason why such a machine was impossible wasn't understood until Rolf Landauer re-conceptualized the paradox in terms of computation. Any Maxwellian demon would need to erase its memory, which requires energy and in the process would offset any energy gained from effortlessly separating its particles [11]. Ideas such as these borrowed from computational complexity may ultimately help physics to resolve the asymmetry of time in the classical and quantum worlds [12].

We haven't yet even really touched upon quantum mechanics where not only is computational complexity informing physics, but physics is revolutionizing computer science itself. In just the last decade quantum computers have gone from science fiction to working models. More sophisticated versions of these machines, or even the discovery that we are incapable of building them, may finally allow us to resolve the dispute between the different versions of quantum mechanics—Schrodinger's, Heisenberg's or Feynman's—all of which make identical predictions but are built on radically different ontologies [6]. Quantum computers may also give us some indication whether the Many Worlds Interpretation is real or a mirage [13]. Computational complexity has already indicated that Bohmian Mechanics is intractable, and thus might give us a clue that hidden variable theories must be wrong [14].

More astounding is the impact the study of quantum mechanics has had on computational complexity itself. The realization that quantum entanglement can be used as a resource for computation in the form of Multi-Prover Interactive Proofs (MIP*) has resulted in an exciting marriage of the two fields of study the most profound of which seems to show that many forms of uncomputability might be overcome with the resource of infinite entanglement.[1] Unfortunately, this is not a resource found in the universe in which we live.

There have by now been so many instances of where the impossibility of Laplacian demons has allowed us to infer aspects of the universe that physicists might want to start thinking about using computational complexity itself as a way to infer constraints in the way they now use the anthropic principle. Any theory that would permit P = NP as in faster than light communication or time travel (closed time like curves), or which gave rise to memory or speed capacities beyond the Bekenstein Bound, could be bracketed as unlikely for that very reason.

---

[1] Ji et al. [15]. An updated version of the above cited paper corrects for an error and shows that *MIP = RE still stands. See: Zhengfeng Ji, Anand Natarajan, Thomas Vidick, John Wright, Henry Yuen "Quantum soundness of the classical low individual degree test" https://arxiv.org/abs/2001. 04383.

Still none of this will answer what is perhaps the ultimate question when it comes to the role of computation in our world, the question that lies at the heart of our discovery of computation itself—is the world at bottom deterministic or random (in the sense of Wheeler's "law without law") [16]? Everything so far would appear to suggest that it is a finely tuned balance of both [17]. What makes the question perhaps impossible to answer is the dilemma of self-reference which is not a problem we can solve. We try to look at the universe from an Archimedean point outside of it and yet are inescapably trapped within, compelled to choose the role we play while living here. If fate exists then here is where it can be found, for we are forced to confront some version of Gödel's Incompleteness or Turing's Halting Problem in the radically different worlds of black holes, quantum measurements, or even when we try to understand what it means to be an agent and a self [18].

References

1. R.C. Hilborn, *Chaos and Nonlinear Dynamics: An Introduction for Scientists and Engineers* (Oxford University Press, New York, 2006)
2. E.N. Lorenz, Deterministic non-periodic flow. J. Atmos. Sci. **20**(2), 130–141 (1963)
3. S.A. Cook, The complexity of theorem-proving procedures, in *Proceedings of the Third Annual ACM Symposium on Theory of Computing—STOC 71* (1971)
4. C.H. Papadimitriou, M. Yannakakis, Shortest paths without a map, in *Lecture Notes in Computer Science. Proceedings of 16th ICALP*, vol. 372 (Springer, 1989), pp. 610–620
5. S. Aaronson, *Computational Complexity and the Anthropic Principle*. Talk at: Stanford Institute for Theoretical Physics (2006). https://www.scottaaronson.com/talks/anthropic.html
6. S. Aaronson, *Lens of Computation on the Sciences*. Institute for the Advancement of Science (2014). https://www.youtube.com/watch?v=hJibtaTmLtU
7. J.D. Bekenstein, Black holes and entropy. Phys. Rev. D. **7**, 2333 (1973)
8. L. Susskind, Computational complexity and black hole horizons. Fortschritte Der Physik **64**, 24 (2016)
9. P. Hayden, J. Preskill, Black holes as mirrors. J. High Energy Phys. 120 (2007)
10. I.H. Kim, E. Tang, J. Preskill, The ghost in the radiation: Robust encodings of the black hole interior (2020). https://arxiv.org/abs/2003.05451
11. R. Landauer, Irreversibility and heat generation in the computing process. IBM J. Res. Develop. **5**(3), 183–191 (1961)
12. J. Thompson et al., Causal asymmetry in a quantum world. Phys. Rev. X **8**, 031013 (2018). Published 18 July
13. G. Jaeger, *Entanglement, Information, and the Interpretation of Quantum Mechanics* (Springer, Berlin, 2009)
14. S. Aaronson, Quantum computing and hidden variables. Phys. Rev. A **71** (2005)
15. Z. Ji, A. Natarajan, T. Vidick, J. Wright, H. Yuen, *MIP = RE (2020). https://arxiv.org/abs/2001.04383
16. J.A. Wheeler "On recognizing 'law without law," Oersted Medal Response at the joint APS AAPT Meeting, New York, 25 January 1983
17. T. Austin, Measure concentration and the weak Pinsker property. Publications Mathématiques De LIHÉS **128**, 1 (2018)
18. D.R. Hofstadter, *Gödel, Escher, Bach: An Eternal Golden Braid* (Basic Books, New York, 1979)

Appendix
List of Winners

First Prizes

Klaas Landsman: Undecidability and indeterminism.[1]

Markus Müller: Undecidability and unpredictability: not limitations, but triumphs of science.

Second Prize

David Wolpert, David Kinney: Noisy Deductive Reasoning: How Humans Construct Math, and How Math Constructs Universes.

Third Prizes

Flavio Del Santo: Indeterminism, causality and information: Has physics ever been deterministic?

Hippolyte Dourdent: A Gödelian Hunch from Quantum Theory.

Tim Palmer: Undecidability, Fractal Geometry and the Unity of Physics.

---

[1] From the Foundational Questions Institute website: https://fqxi.org/community/essay/winners/2020.1.

Fourth Prizes

Rade Vuckovac: Unpredictability and Randomness.

Ian Durham: Why is the universe comprehensible?

Prize for an Interesting Literary Discourse

Rick Searle: Computational Complexity as Anthropic Principle.

Prize for a Creative Approach to the Problem

Jochen Szangolies: Epistemic Horizons: This Sentence is $1/\sqrt{2}(|\text{True}\rangle + |\text{False}\rangle)$.

9783030703530